A Practical Guide to

Developmental Biology

Melissa Gibbs

Department of Biology, Stetson University, Florida

OXFORD
UNIVERSITY PRESS

Great Clarendon Street, Oxford OX2 6DP

Oxford University Press is a department of the University of Oxford.
It furthers the University's objective of excellence in research, scholarship,
and education by publishing worldwide in

Oxford New York

Auckland Bangkok Buenos Aires Cape Town Chennai
Dar es Salaam Delhi Hong Kong Istanbul Karachi Kolkata
Kuala Lumpur Madrid Melbourne Mexico City Mumbai Nairobi
São Paulo Shanghai Taipei Tokyo Toronto

Oxford is a registered trade mark of Oxford University Press
in the UK and in certain other countries

Published in the United States
by Oxford University Press Inc., New York

First published 2003

British Library Cataloguing in Publication Data
Data available

Library of Congress Cataloging in Publication Data
Data available

ISBN 0199249717

1 3 5 7 9 10 8 6 4 2

Typeset by SNP Best-set Typesetter Ltd., Hong Kong
Printed in Great Britain
on acid-free paper by
Ashford Colour Press Ltd.
Gosport, Hants.

A Practical Guide to Developmental Biology

PREFACE

This laboratory manual is designed to compliment the chapter sequence of most developmental biology textbooks and to expose students to the widest variety of experimental animals possible. Care has been taken to present experimental protocols in as straightforward a manner as possible and to include extensive lists of sources and recipes for the instructor. The exercises are a combination of classic embryology, slide work, and cutting-edge experimentation (albeit with budgets in mind). The introductory lab session, during the first 2 weeks, is devoted to acquainting and instructing students in animal experimentation ethics, clean embryo techniques, staging of various embryos, and how to successfully handle a variety of embryo types. This first chapter will also serve as a valuable reference for later lab exercises.

Preparing for a laboratory session

Anyone who has taught developmental biology knows that when you teach it, considerable planning and flexibility is required. It takes time to order in chemicals, to mix up solutions, and, most importantly, to order and receive experimental animals. For example, it may take up to 3 weeks to receive chick embryos or 2 weeks for urchins. Although it doesn't take long to get a flask of fruit flies shipped, you will still need a week or two to generate a large enough supply of egg-laying adults, and then a few days more for the eggs to develop into usable larvae. Also, embryo vendors rarely deliver on Mondays and may only ship or deliver on a few select days. A final complication which must be considered is that embryo availability and viability varies with the seasons. Chick embryos, for example, have a 40–45% viability rate in the fall and a 65–75% viability rate in the spring.

Appendices are included on the husbandry of all the organisms you will be using and where and when you can purchase them. Other appendices include a list of all chemicals and slides needed, as well as their sources. A complete list of recipes for all solutions and hand-made materials is also included.

Conducting the laboratory session

In the second heading of each chapter, immediately following the title of the exercise, is the amount of time required to complete the exercise. Although most labs meet once a week, the nature of developmental biology labs necessitates some out-of-class observation and manipulation time. For example, salamander embryos exposed to a mutagen like retinoic acid will need to be sketched every day and have their rearing solution changed every other day. It is vital, therefore, that students have access to their experiments outside of the lab period.

Ideally, each lab exercise will have been preceded by a lecture on that topic, but lab/lecture coordination is rarely ideal and exercises may need to be juggled due to the availability of experimental animals. A few paragraphs of background material are provided for each exercise to introduce the student to the issues at hand or to serve as a brief review. A list of solutions and equipment is provided for each exercise and should be reviewed by both the instructor and the student. The experimental protocols are laid out in a step-by-step manner, thereby making it easier for the student to follow instructions and plan their time. Troubleshooting hints and chemical warnings (if needed) are provided for each exercise as well.

Analyzing the laboratory session

Lab questions are provided at the end of each exercise to help the student think about their experimental results. With the exception of the introductory lab, all of these exercises lend themselves to written lab reports. Students should be encouraged to keep detailed lab notebooks, which the instructor may wish to assess at least once during the semester.

Acknowledgments

This manual would not have been possible without the work of the many developmental biologists who preceded me. Major inspiration and influence came from the papers and manuals of Viktor Hamburger, Thomas Hunt Morgan, Gavin De Beer, Julian Huxley, Johannes Holtfreter, Roberts Rugh, Gary Schoenwulf (Exercise 9), Mary Tyler, Leland Johnson, Laura Keller, John Evans and Thomas Keller, Yolanda Cruz, and Rodney Scott (Exercise 5).

Glenn Northcutt and members of his lab at the Scripps Institution of Oceanography indoctrinated me into the weird and wonderful realm of developmental biology, but my parents inspired my love of science first.

Many thanks are due to my students at Stetson University, who have been the guinea-pigs for the exercises in this manual. Thanks are also due to the following instructors who kindly gave their advice during the development of this manual: Dr Nicole Bournais, California State University, San Bernardino, USA; Professor Ana Campos, McMaster University, Canada; Professor Julie Emerson, Saint Louis University, USA; Professor Clive Evans, University of Auckland, New Zealand; Dr Neil Haave, Augustana College, Canada; Dr Tony Jelsma, Dordt College, USA; Dr Christopher Rose, James Madison University, USA; Dr Alyssa Perez-Edwards, Duke University, USA; Professor Wayne Wiens, Bethel College, USA.

Finally, I thank my editor, Jonathan Crowe, who has deftly guided me through the grind and intricacies of book writing.

MAG

September 2003

CONTENTS

PART 1

The Experiments

INTRODUCTION TO DEVELOPMENTAL BIOLOGY

1

Embryo protocols, ethics, and model systems

2 weeks

Before you can begin to conduct experiments on developing embryos, you must familiarize yourself with the general protocols under which embryos are handled, the ethical considerations of using living organisms in a laboratory setting, and the basic developmental patterns of your model systems. The following experiments and exercises are designed to do just that over a 2-week period. In addition, your instructor may choose to show you a video of time-lapse microscopy of normal development. There is a short list of available titles at the end of this chapter.

The general embryo protocols listed below should remain in the back of your mind whenever you work with embryos. If these protocols are not followed, your experiments will either be more difficult than they should, or will fail because you have allowed infectious agents to enter your embryo chambers. After you have carefully read through the embryo protocols, you will read brief statements about the ethics of working with living organisms and why we tend to use just a few model systems in our investigations. Then, finally, the animals themselves—this laboratory exercise is designed to familiarize you with a number of basic techniques and allow your fingers to get used to manipulating very small, delicate organisms.

Most of the model systems we will be using have some sort of embryonic protective shield (jelly coat, shell, chorion, etc.). Since we may be removing that shield, additional measure must, however, be taken to protect the embryos from injury or infection.

General embryo protocols

1. Always use sterile techniques (sterilize instruments and vessels with 70% alcohol or by boiling. If you drop an instrument or touch the tips with your fingers, sterilize it again!).
2. Do not breathe on the embryos or leave your mouth open while you are leaning over an open Petri dish.
3. Use glassware that is new or designated for embryos only. Noxious chemicals can be absorbed into the glass, and then be released back into your experiment, having a deleterious effect on developing embryos. Sterile, plastic Petri dishes are great, but expensive. Glass Petri dishes, on the other hand, can be used over and over again if rinsed 10 times in tap water, 10 times in distilled water, and then boiled for 5 minutes to sterilize.

4. You will be using extremely fine-tipped forceps that will become bent (and therefore of much less, or no, use to you) if handled roughly or dropped. Please be very careful with them!
5. Because you will often have several different experimental conditions underway at a time, make sure everything is correctly labeled. Labeling should be done on the bottom half of the Petri dish (in case the top comes off) around the periphery of the bottom of the dish (so you have a relatively unobstructed view of your embryos), and written in small letters with a marking pen (again to give you an unobstructed view of your embryo). Don't forget to add your name!
6. Refer to embryos by their stage. When dealing with embryos, we are much more interested in their developmental achievements than exactly how old they are. Therefore, when discussing embryos, we refer to their developmental stage. For example, most embryos go through some sort of blastula, gastrula, and neurula stages before they begin to develop features like eyes, ears, gills, and limb buds. Embryo stages can be either descriptive or numeric (e.g. a midneurula axolotl embryo is stage 16). Refer to Figs 1.1 and 1.7–1.13 for fruit fly, amphibian, zebrafish, and chick stages.

Ethics

The ethics behind using living material in a lab must always be considered. First and foremost in the mind of any experimenter should be whether a procedure will cause an animal any unnecessary distress or pain. Strict government guidelines for animal use are currently in place for any researcher seeking government funding (National Institutes of Health *Guide for the care and use of laboratory animals*); and, although these guidelines were written for US government-funded scientists who use terrestrial warm-blooded animals, they should (and are intended to) be considered by anyone using living organisms in their lab. Many of the organisms used in the experiments detailed in this lab manual will not survive to adulthood due to their general fragility. Whenever organisms are no longer needed, they should be anesthetized, euthanized, and properly disposed of.

Model systems

We use model systems in science for a variety of reasons: in developmental biology, our model systems are chosen for their ready availability, short generation times, large egg or embryo size, and well-described developmental stages. During your introduction to developmental biology, you will become familiar with, among others, the development of fruit flies (*Drosophila*), killifish, aquatic salamanders (axolotls), chicks, bean plants, and water fern. More expensive, time-consuming model systems (sea urchins, live chick embryos, etc.) will be introduced at a later time.

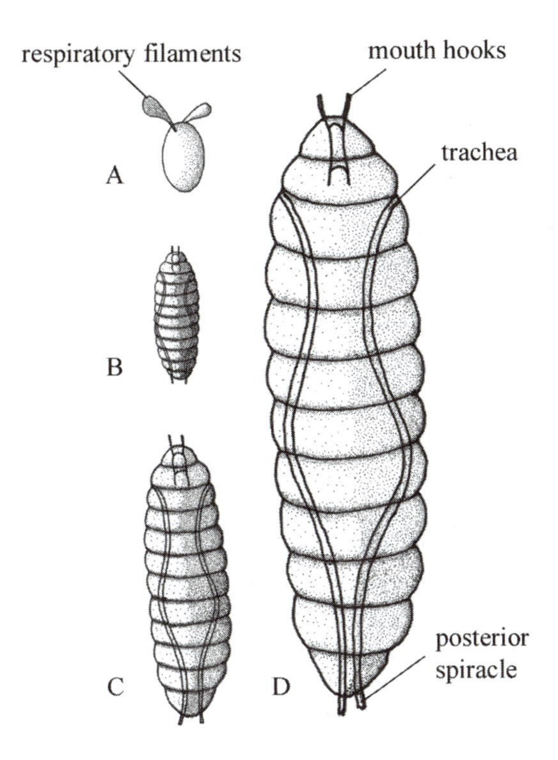

Fig. 1.1 Developmental stages of *Drosophila*, illustrating major external anatomical features. (A) embryo; (B) 1st-instar larva; (C) 2nd-instar larva; (D) 3rd-instar larva. Scale bar = 0.5 mm

Drosophila

The fruit fly (*Drosophila*) is a very common experimental animal for genetics and developmental biology, due to its short generation time (about 2 weeks), ease of maintaining colonies, well-known genetics, and wide variety of known mutants. When the most common species, *D. melanogaster*, is incubated at 25 °C, it lays eggs at 2 days of age. Eggs hatch in 24 hours, and the larvae go through two additional molts at 24-hour intervals (Fig. 1.1). The 3rd instar larva pupates after 96 hours, and hatches 4 days later. These times will vary somewhat with the species and incubation temperatures. *Drosophila virilis* is a large species of fruit fly that is preferred for studies requiring the manipulation of larvae.

Although *D. virilis* is considerably larger than *D. melanogaster*, the dissection and manipulation of fly embryos is difficult. Not only are you dissecting larvae that are less than 5 mm long, but their entire body is clear to white in color, making it difficult to distinguish one tissue type from another. Conducting today's experiment will give you the necessary experience to conduct later *Drosophila* experiments with considerably more ease.

Drosophila have very large multi-stranded chromosomes (polytene chromosomes) in their salivary glands. These large chromosome complexes contain 1000 copies of each individual chromosome and are visible under the light microscope (Fig. 1.2(A)). The polytene chromosomes are so large that you can actually see the puffed out areas (Fig. 1.2(B)) where active transcription is taking place (remember that to be transcribed, the DNA must unravel the chromosome packaging)! Today, after you play around with the different stages of the *Drosophila* life cycle, you will be making chromosome squashes to see the polytene chromosomes. Incubating (heat-shocking) larvae at 37 °C for 30 minutes will induce puffing in different sites, so be sure to make a squash of a heat-shocked larva for comparison. While you are dissecting out the salivary glands, keep an eye open for the imaginal discs. These discs, which will give rise

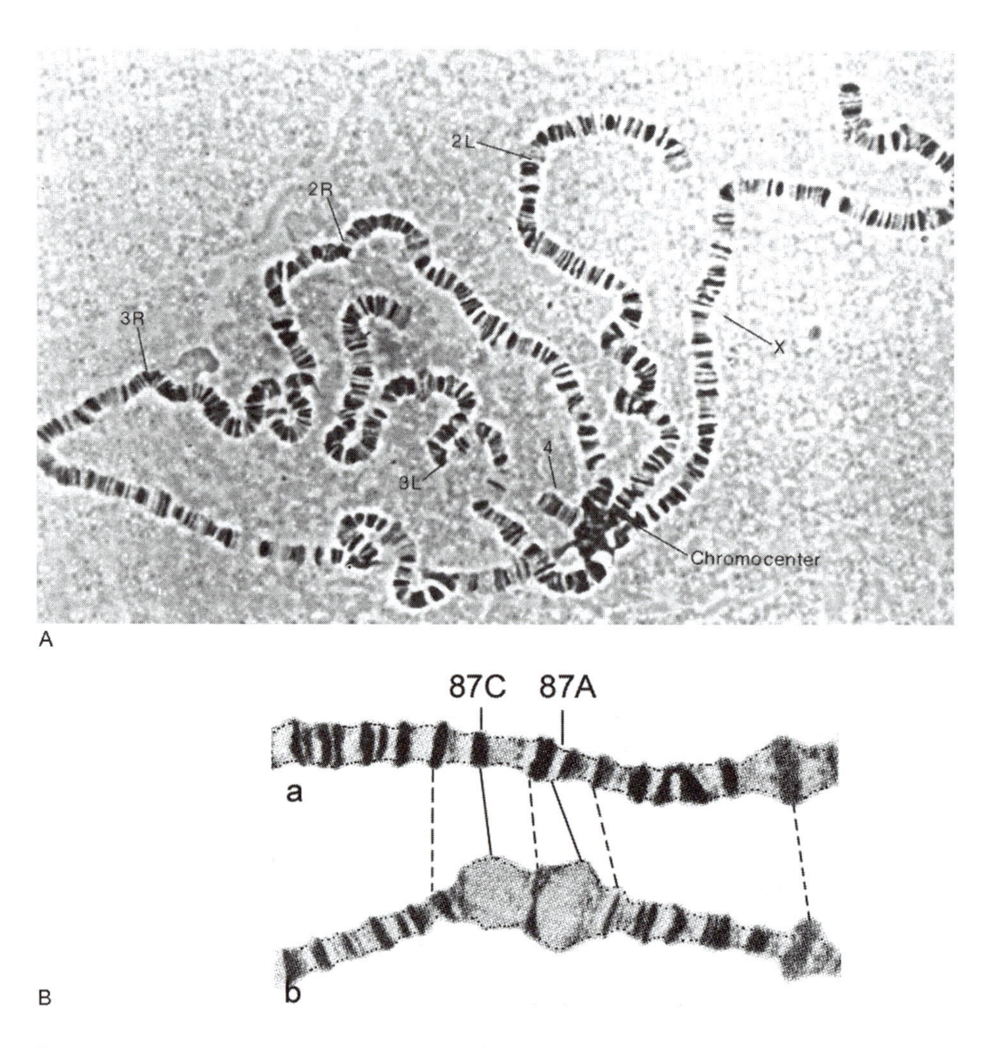

Fig. 1.2 (A) Polytene chromosomes from the salivary glands of *Drosophila* after a chromosome squash; (B) control (top) and heat-shocked (bottom) polytene chromosomes. (Reprinted from *Cell*, vol. 17, M. Ashburner and J. J. Bonner, The induction of gene activity in *Drosophila* by heat shock, pp. 241–54, 1979, with permission from Elsevier Science.)

to leg, wings, and other adult structures, look like coiled mats, are relatively small, and are concentrated in the anterior one-third of the larva (Fig. 1.3).

Drosophila salivary gland extraction protocol

1. Examine, sketch, and take notes on embryos, all three larval stages, early and late pupae, and adults. The embryos, larvae, and pupae will be alive, but the adults will be dead.
2. Remove a large, third-instar larva (these should be crawling up the side of the culture bottle) with a small paintbrush and place it in a drop of Insect saline on a slide.
3. While looking through the dissecting microscope, use your forceps to grip the head (position your forceps directly behind the mouth hooks in the first body segment). Grab the body at its widest point with your second pair of forceps (Fig. 1.4).
4. Gently pull the body away from the head. You should be left with the first segment of the head, the brain, salivary glands, and some fatty tissue.
5. Gently tease away the fatty tissue, brain, and mouth hooks. Be sure to keep the salivary glands moist with Insect saline.
6. Draw the saline off the glands using twists of lab tissue, and immediately add a drop or two of acetic acid to fix the glands. Leave the fixative on for at least 1 minute.
7. Draw off the acetic acid and replace with aceto-orcein stain. Allow the glands to stain for 15 minutes. Keep an eye on the stain to make sure it doesn't dry out.
8. Place a coverslip over the stained glands, being careful to avoid air bubbles around the glands.
9. Fold a lab tissue into eighths, place it on the coverslip, and press **very** firmly with your thumb or a pencil eraser. Do not allow the coverslip to move much relative to the slide.
10. Examine the chromosomes at 10× and 40× magnifications.
11. If you didn't get a good squash, repeat the process until you do.
12. Make a chromosome squash of a heat-shocked larva.

Axolotl

The axolotl (*Ambystoma mexicanum*) is an aquatic salamander that retains its gills as an adult (a paedomorphic feature). Axolotls are very popular experimental animals because their embryos are relatively large (3–4 mm diameter) and are commercially available at nearly all stages (from Indiana University's axolotl colony) all year round. Albino embryos are also available. You will be using axolotl embryos in several exercises this semester.

At 29 °C, axolotl embryos will reach early blastula stage at 16 hours, early gastrula at 26 hours, early neurula at 50.3 hours, gill- and tailbud stages at 73 hours (see Table 1.1), gills at 130

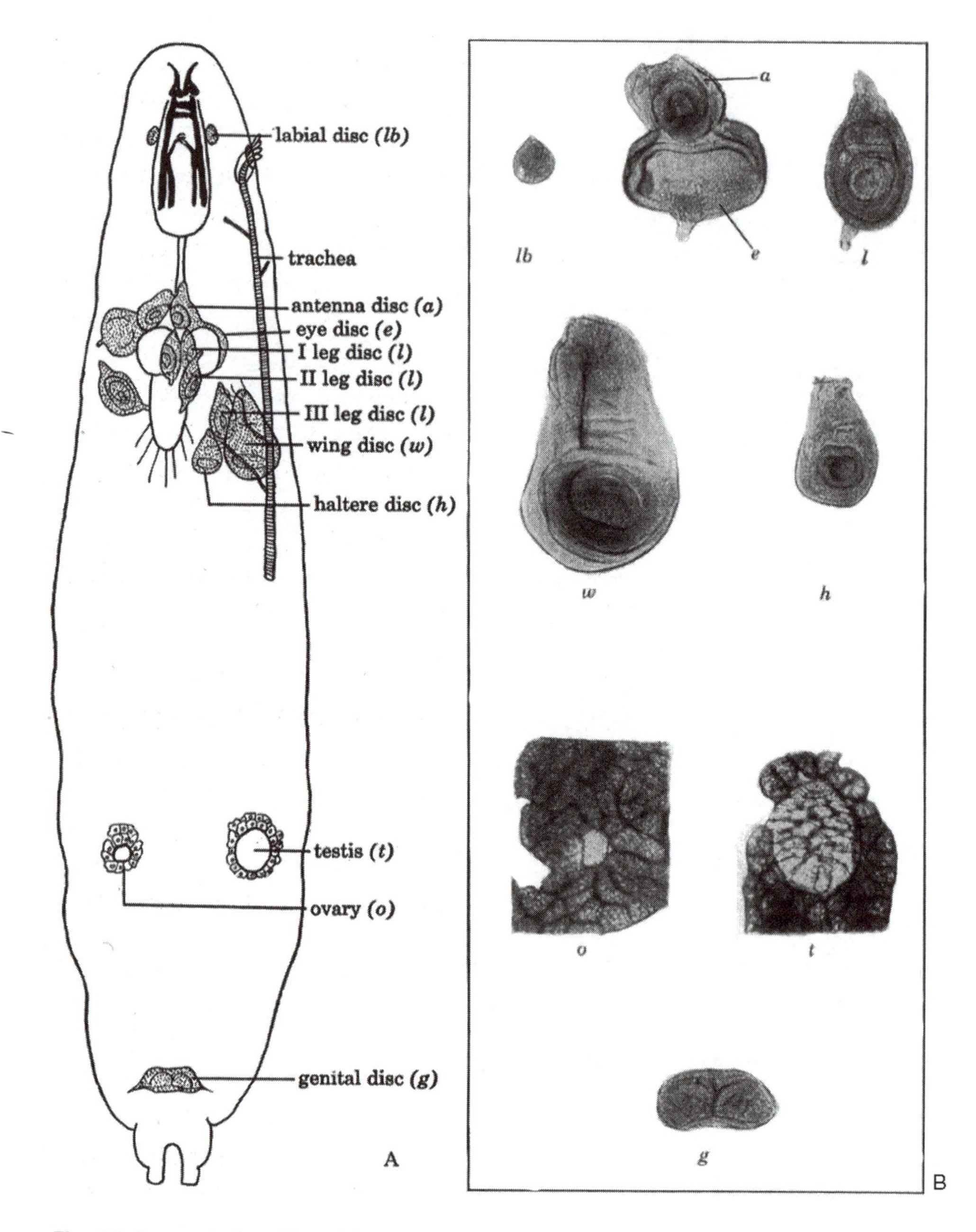

Fig. 1.3 Anatomical position of *Drosophila* imaginal discs. Notice how most discs are clustered together near the brain. (From M. Demerec, *The Biology of* Drosophila, 1950.)

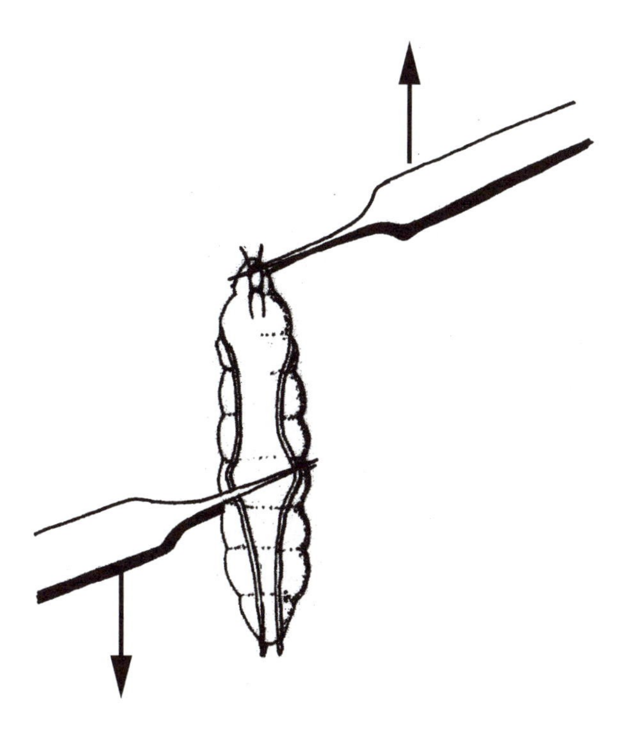

Fig. 1.4 Technique for dissecting out imaginal discs. Grasp the larva behind its mouth hooks and at the mid-body to pull apart the head and body to reveal salivary glands and imaginal discs.

hours, and they will hatch at about 300 hours (about 12 days) (Armstrong and Malacinski 1989) (Figs 1.5 and 1.6). At room temperature, this time line will be extended to about 2 weeks.

Like all amphibians, axolotl embryos are protected from infection and mechanical injury by a tough jelly coat (egg capsule) that must be removed for most experimental procedures. To maximize the survival of your experimental animals, the outer jelly coat must be sterilized before dejellying takes place. You will find this jelly coat quite difficult to penetrate at first; however, it is important to become adept at the dejellying technique before you attempt to begin any experiments. The inner vitelline membrane is left intact.

Disinfection protocol

1. Place embryos in a 0.01% formalin solution for 10 minutes. This treatment kills any fungi or bacteria adhering to the outside of the jelly coat.
2. Sterilize two pairs of forceps by soaking them in 70% ethyl alcohol (ethanol) for 10 minutes.
3. Using a clean plastic pipette (slightly cut down so the opening is larger than the embryo), transfer the embryos from the formalin solution into sterile Rearing solution (RS). Rearing solution is an ionically balanced solution containing antibiotics that will

help prevent subsequent infections. Make sure that a minimum of the formalin solution is transferred with the embryos.

Table 1.1 Stages of normal axolotl (*Ambystoma mexicanum*) development at 29 °C

Stage no.	Time from 1st cleavage (h)	Features
1	—	Freshly laid, fertilized
2	0	Cleavage
3	2.4	4 cells
4	4.12	8 cells
5	5.22	16 cells
6	6.45	32 cells
7	8.26	64 cells
8	16.06	Early blastula
9	21.28	Late blastula
10	26	Early gastrula, blastopore lip starts to become visible
11	38.3	Middle gastrula, dorsal lip of blastopore forms semicircle
12	47.3	Late gastrula, blastopore is circular
13	50.3	Early neurula, neural folds begin to appear
14	58.15	Broad neural plate
15	59.5	Shield-shaped neural plate
16	63	Middle neurula, neural folds begin to rise and plate begins to sink
17	64.3	Late neurula, folds continue to rise and plate deepens and narrows near the head
18	66	Neural folds closing
19	69	Neural folds touching
20	70.3	Neural folds fusing
21	72	Neural folds fused
22	73	Gill region, pronephros and tailbud distinct
23	74	Ear primordium visible
24	80	Ear pit and somites distinct

Source: Armstrong and Malacinski (1989).

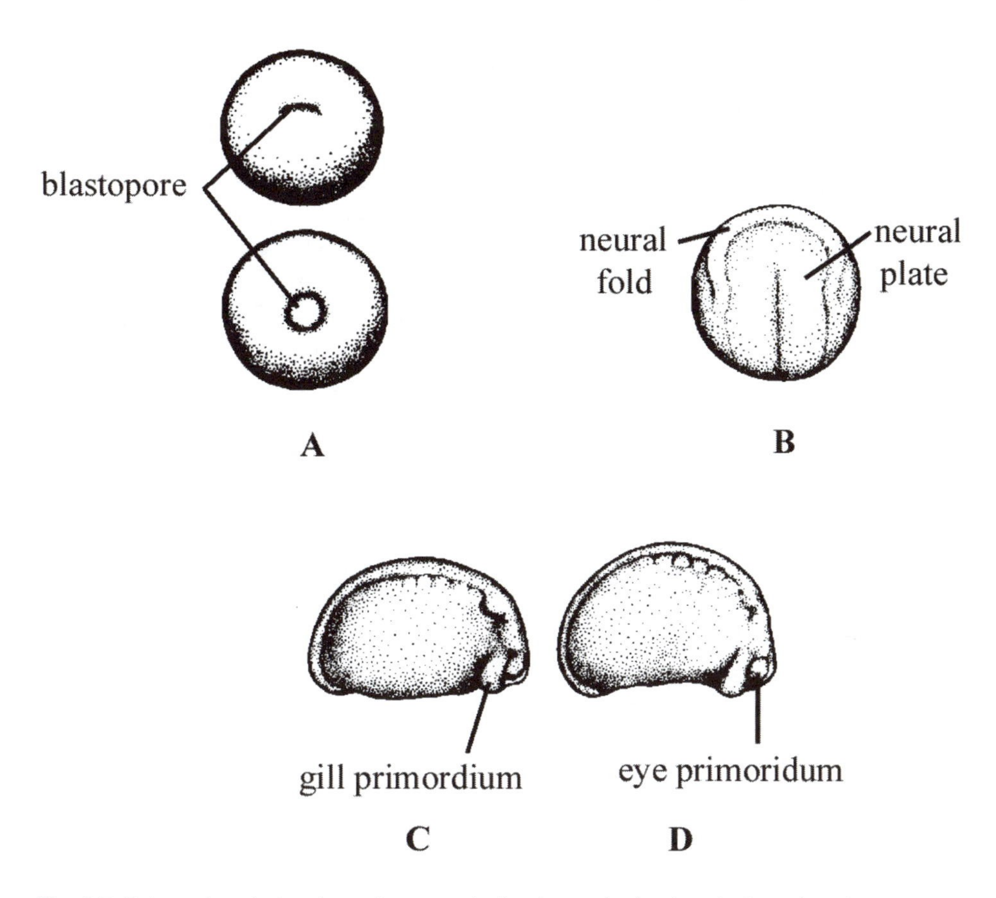

Fig. 1.5 Selected axolotl embryonic stages indicating major landmarks in early to late embryos. (A) stage 10–12; (B) stage 16; (C) stage 20; (D) stage 21. (Reprinted from *Developmental biology of the axolotl* by John B. Armstrong and George M. Malacinski, copyright 1989 by Oxford University Press, Inc. Used by permission of Oxford University Press, Inc.)

To dejelly the embryos, you need a dissecting microscope, a light source (fiber-optic preferred), two pairs of sterile forceps, and your dish of embryos. You must dejelly under the microscope—if you think you can see it better without the microscope, you are mistaken; readjust your microscope. This technique is one of the more difficult things you will be learning. Before dejellying the embryos, determine their developmental stage using Fig. 1.6.

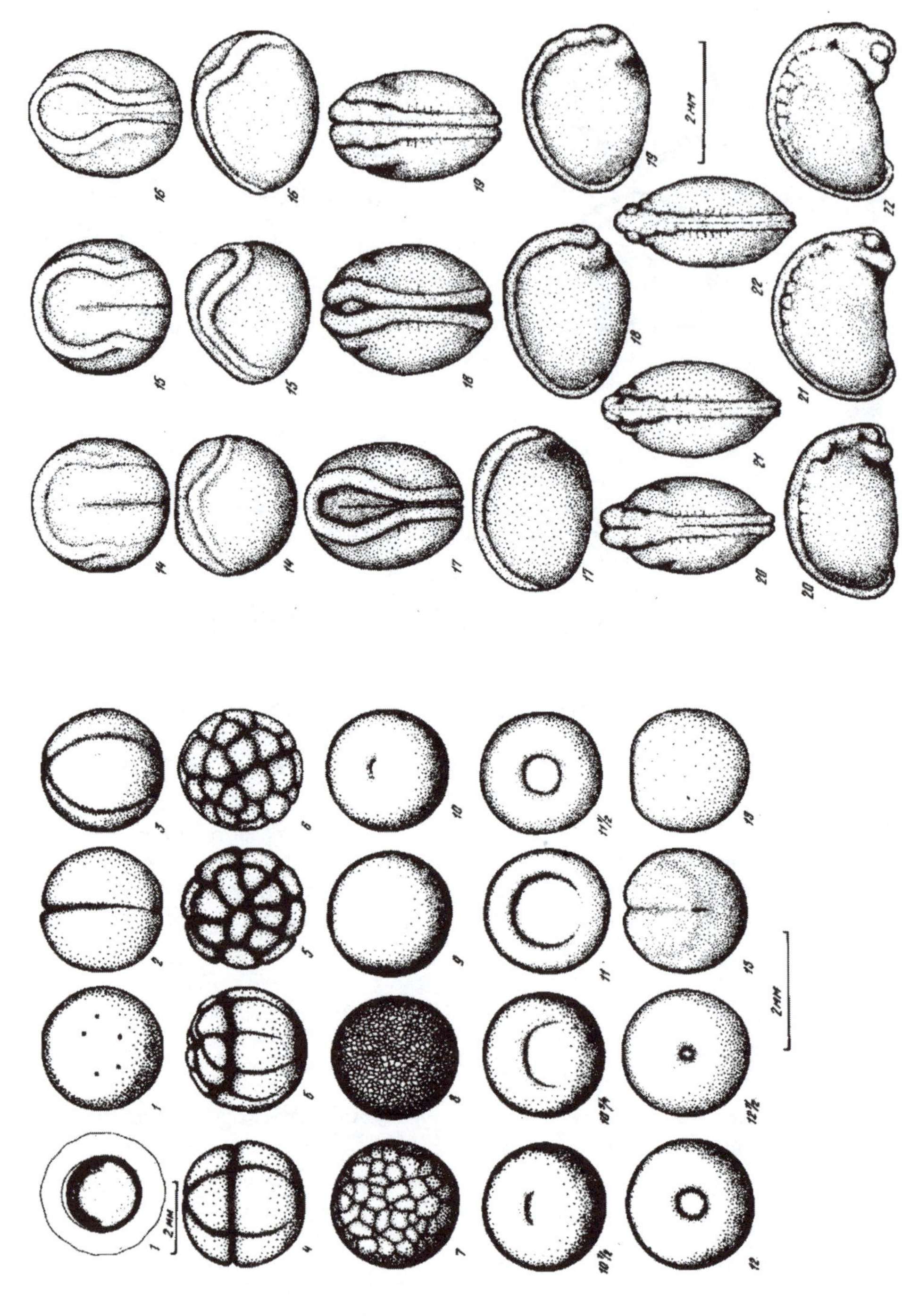

Fig. 1.6 Normal stages of axolotl development. See Table 1.1 for details. (Reprinted from *Developmental biology of the axolotl* by John B. Armstrong and George M. Malacinski, copyright 1989 by Oxford University Press, Inc. Used by permission of Oxford University Press, Inc.)

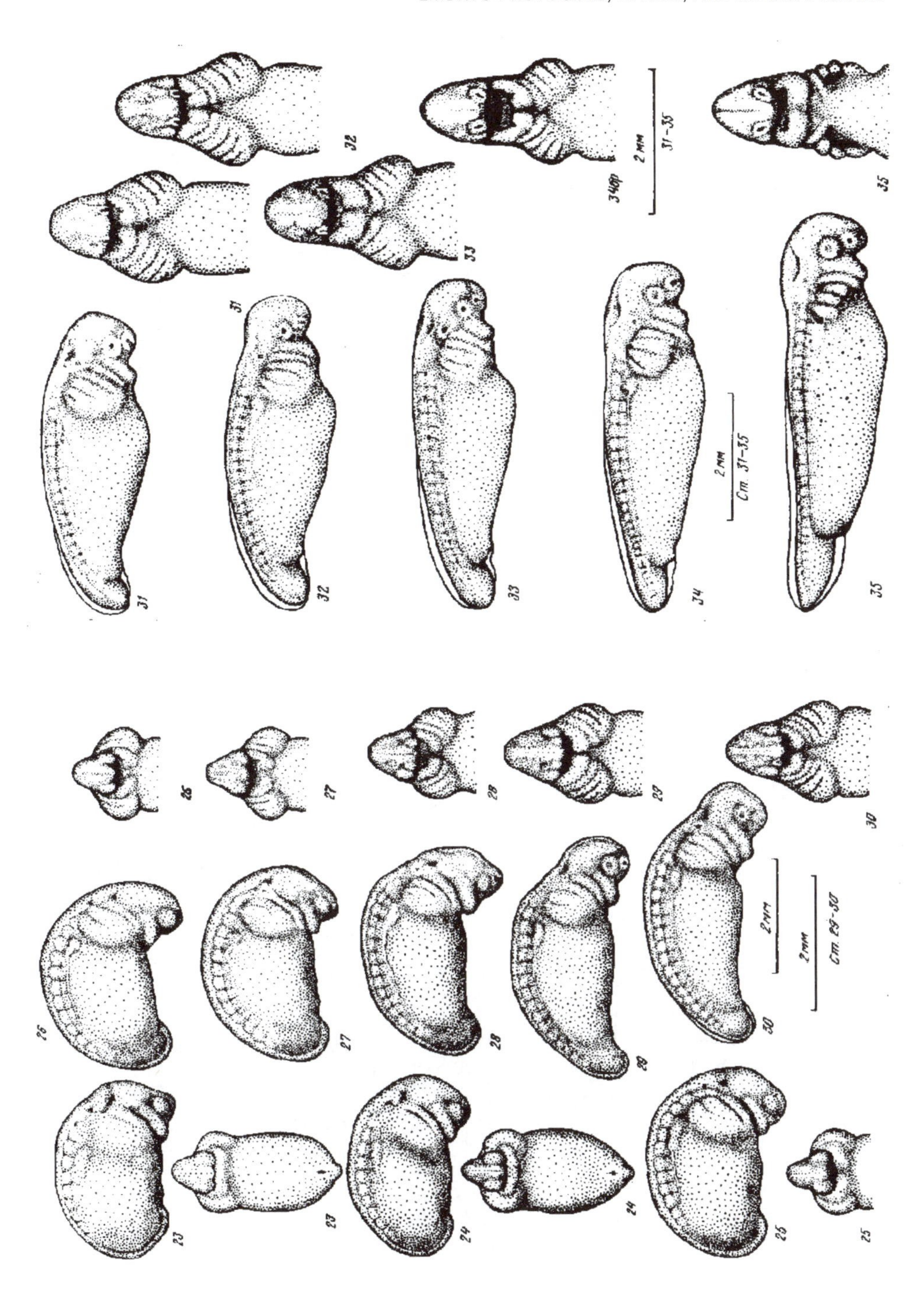

Fig. 1.6 Continued

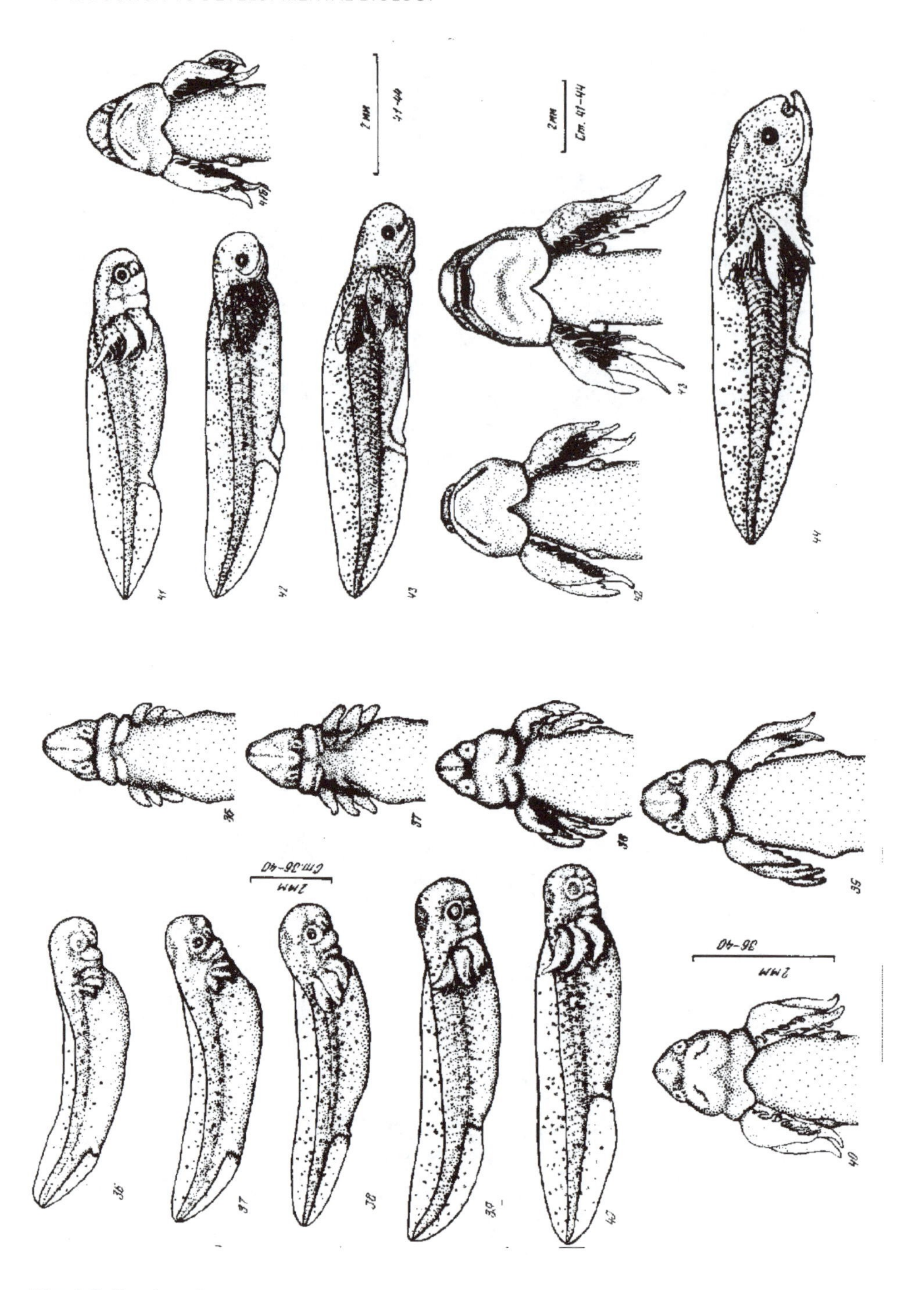

Fig. 1.6 Continued

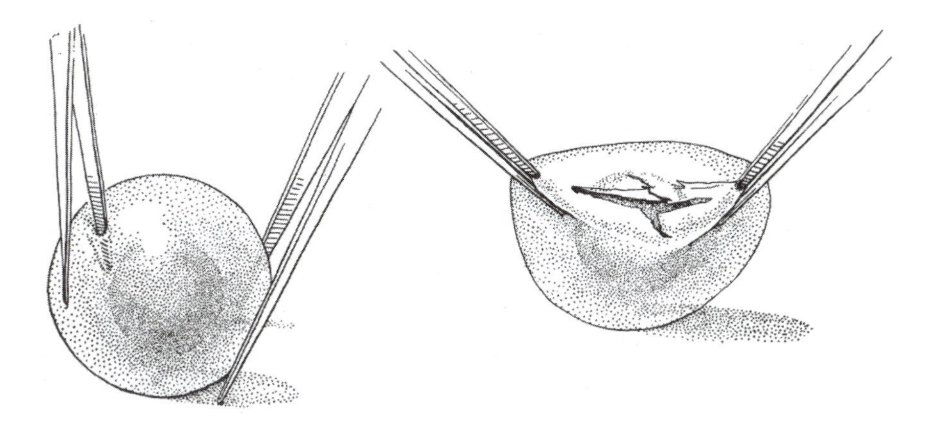

Fig. 1.7 Technique for dejellying an amphibian embryo. (Reprinted from V. Hamburger, *A manual of experimental embryology*, The University of Chicago Press, 1960, with permission of The University of Chicago Press.)

Dejellying protocol

1. Grasp the jelly coat with one pair of forceps. The jelly coat is very slippery, so it may take you several attempts to get a good grip. It can be helpful to use the second pair of forceps to hold the embryo in position while grasping with the other pair (Fig. 1.7).
2. Once you have a good grip on the jelly coat, use the second pair of forceps to poke a hole through the jelly coat. Be careful not to puncture the embryo, and maintain your grip on the jelly coat with both forceps.
3. Pull the jelly coat apart with your forceps, freeing the embryo.
4. Using a cut-down disposable pipette, transfer the dejellied embryo to a fresh Petri dish containing RS.
5. Did the developmental stage have any impact on how easy it was to dejelly the embryo or how well the embryo tolerated the dejellying process?

Once the embryo is free of the protective jelly coat, it is prone to both mechanical and bacterial damage. The antibiotics in the RS take care of most bacterial threats, but **you** must prevent mechanical damage. When you transfer embryos from one dish to another, you must be very careful to introduce the embryo by submerging the pipette tip and gently ejecting the embryo into the RS. If you drop the embryo into the dish, the surface tension of the RS can cause the embryo to explode! When rearing axolotl embryos, no more than five should be placed in a 15 × 30 mm Petri dish (this prevents oxygen depletion and reduces infections). The RS should be replenished every 2–3 days, by siphoning out half of the old solution and then adding fresh RS.

Fish

Killifish are not widely used for studying developmental biology (zebrafish and medaka are much more common), but their embryos do have a unique feature (the ability to enter diapause) that will allow you to actually watch larval fish hatch! Diapause is a period of suspended animation that allows the embryo to survive seasonal droughts that kill adult fish. Killifish cysts are commercially available (Ward's Scientific) and will become active within 5 minutes of hydration. Hatching can occur as early as 15 minutes after hydration, although 0.5–1 hour is more common.

Zebrafish (*Danio rerio*) are currently a popular developmental biology model system. They are readily available from pet stores, produce eggs on a regular basis, and develop rapidly. Obtain a few newly laid zebrafish eggs and observe them throughout the lab period and the following 2–3 days. Since these embryos divide about every 30 minutes and are transparent, they will give you one of the best 'views' of development of any model system (Fig. 1.8).

Chick

Largely due to their great size and availability, chick embryos (of *Gallus domesticus*) have been the subject of developmental biology studies for thousands of years, making their study a vital part of any developmental biology course. Chick embryos are not as resistant to experimental intervention as those of invertebrates and lower vertebrates, but they are invaluable for observing developmental processes. Fertilized eggs are laid at a stage that corresponds to blastula, and within 16 hours (at 38.5 °C) will begin to gastrulate. Major anatomical structures (head, brain, heart, somites, tail, etc.) are clearly visible within 48 hours of laying, and hatching typically occurs 21 days after hatching (Table 1.2 and Figs 1.9–1.13).

During today's lab session, you should familiarize yourself with prepared slides containing wholemount embryos (i.e. the entire embryo has been stained and mounted on a slide) at 18, 24, 36, 48, 72, and 96 hours after fertilization. You will examine these slides again, along with live chick embryos, during a later lab period.

You will also have time to take a close look at the embryo's protective layer, the shell. The shell is considerably more complex than the jelly coat of amphibians; besides a jelly-like coating (albumin), the embryo is protected from mechanical damage and desiccation by a $CaCO_3$ shell. The shell and its underlying membranes are permeable to liquids and gases, so allowing the developing embryo to maintain osmotic equilibrium. Remove a small piece of the shell membrane, mount on a slide, stain with methyl blue, and observe the arrangement of fibers that give this membrane such strength.

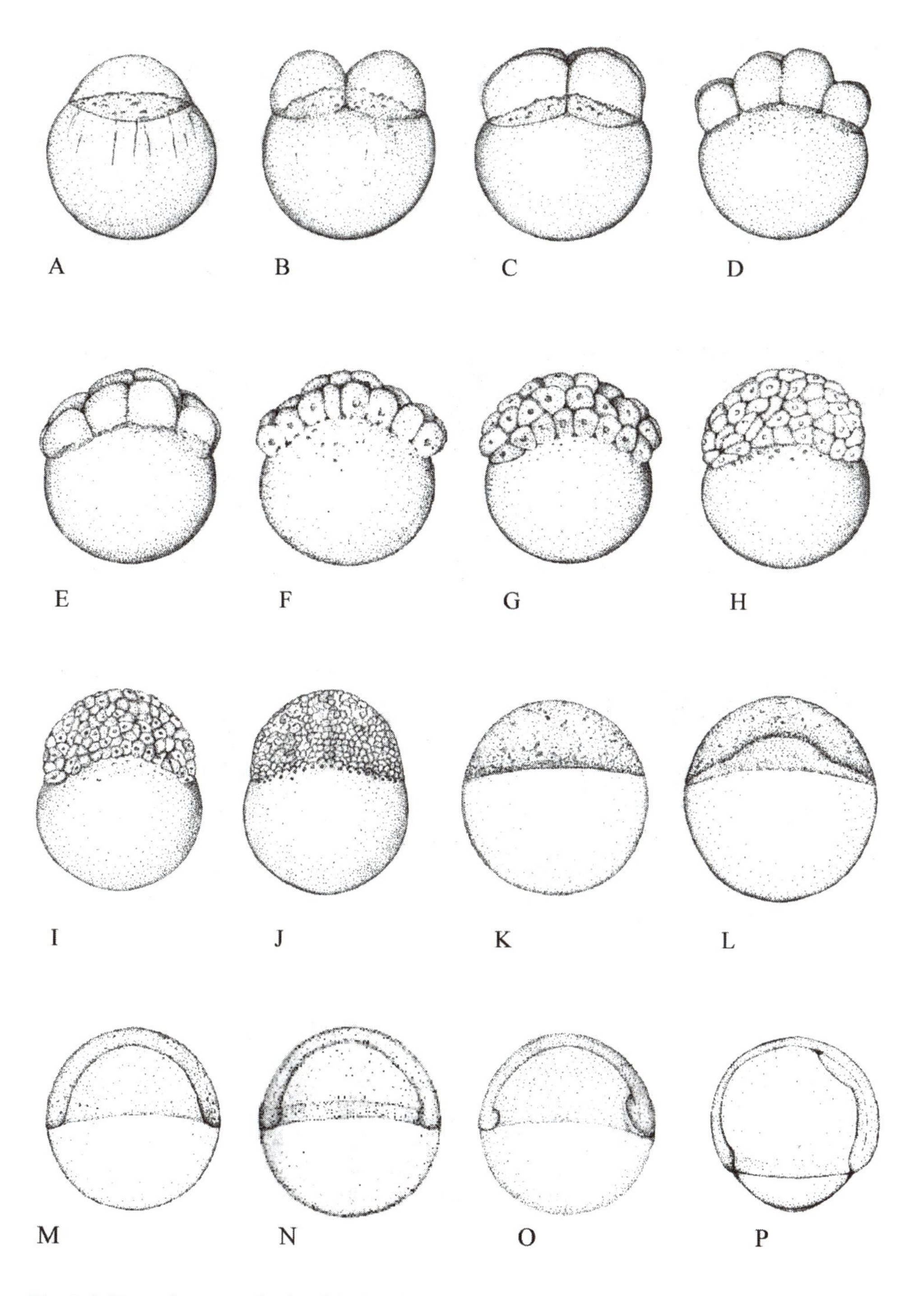

Fig. 1.8 Normal stages of zebrafish development. (A) 1-cell; (B) 2-cell; (C) 4-cell; (D) 8-cell; (E) 16-cell; (F) 32-cell; (G) 64-cell; (H) 128-cell; (I) 256-cell; (J) 1000-cell; (K) sphere; (L) 30% epiboly; (M) 50% epiboly; (N) germ ring; (O) shield; (P) 8 hours; (Q) 9 hours; (R) 10 hours; (S) 11 hours; (T) 12 hours; (U) 14 hours; (V) 16 hours; (W) 18 hours; (X) 19.5 hours; (Y) 22 hours; (Z) 25 hours; (AA) 31 hours; (AB) 42 hours. (Reprinted from *Developmental Dynamics*, vol. 203, C. B. Kimmel, W. W. Ballard, S. R. Kimmel, B. Ullmann, and T. F. Schilling, Stages of embryonic development of the zebrafish, pp. 253–310, 1995, with the permission of Wiley-Liss, Inc., a subsidiary of John Wiley & Sons, Inc.)

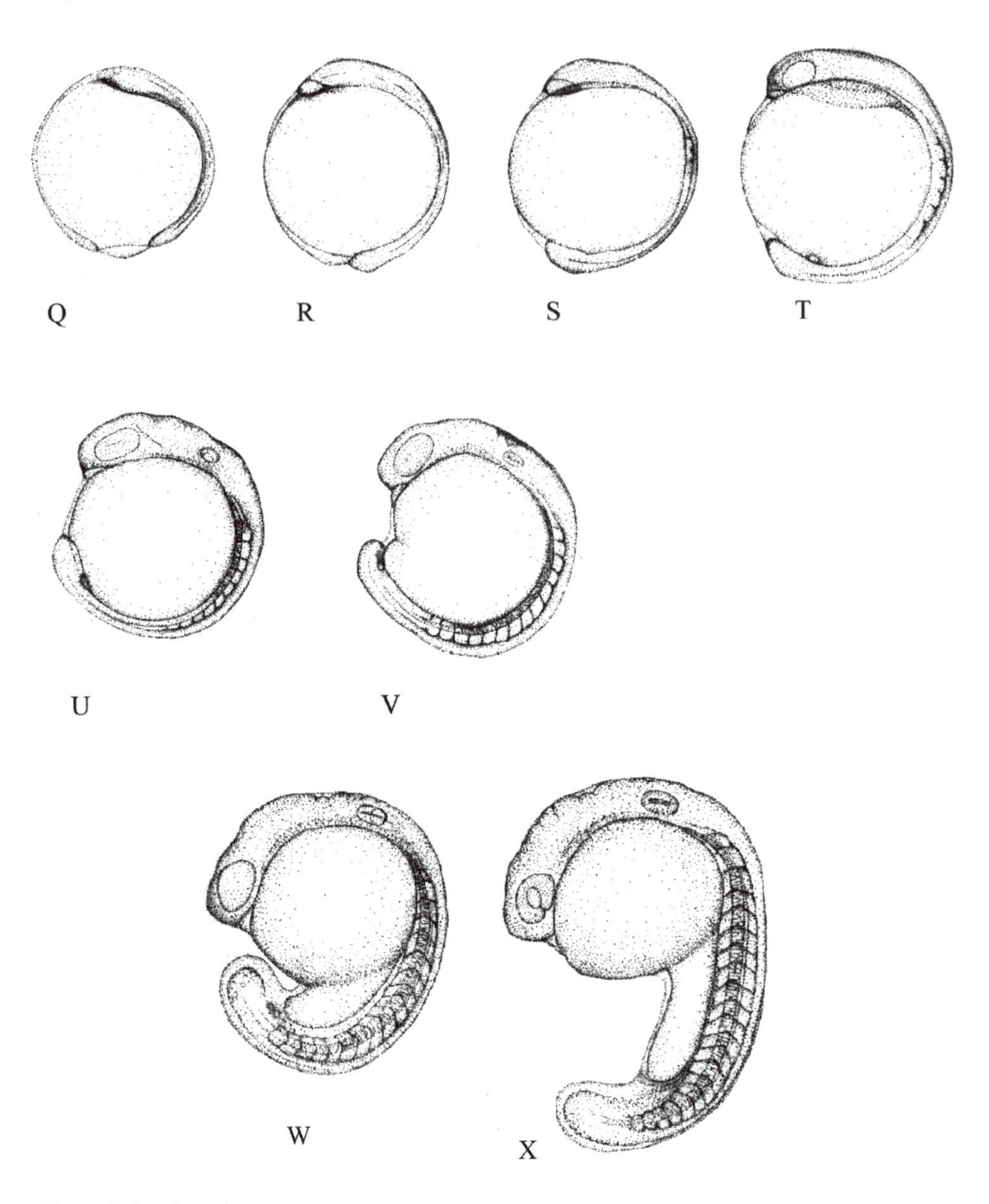

Fig. 1.8 Continued

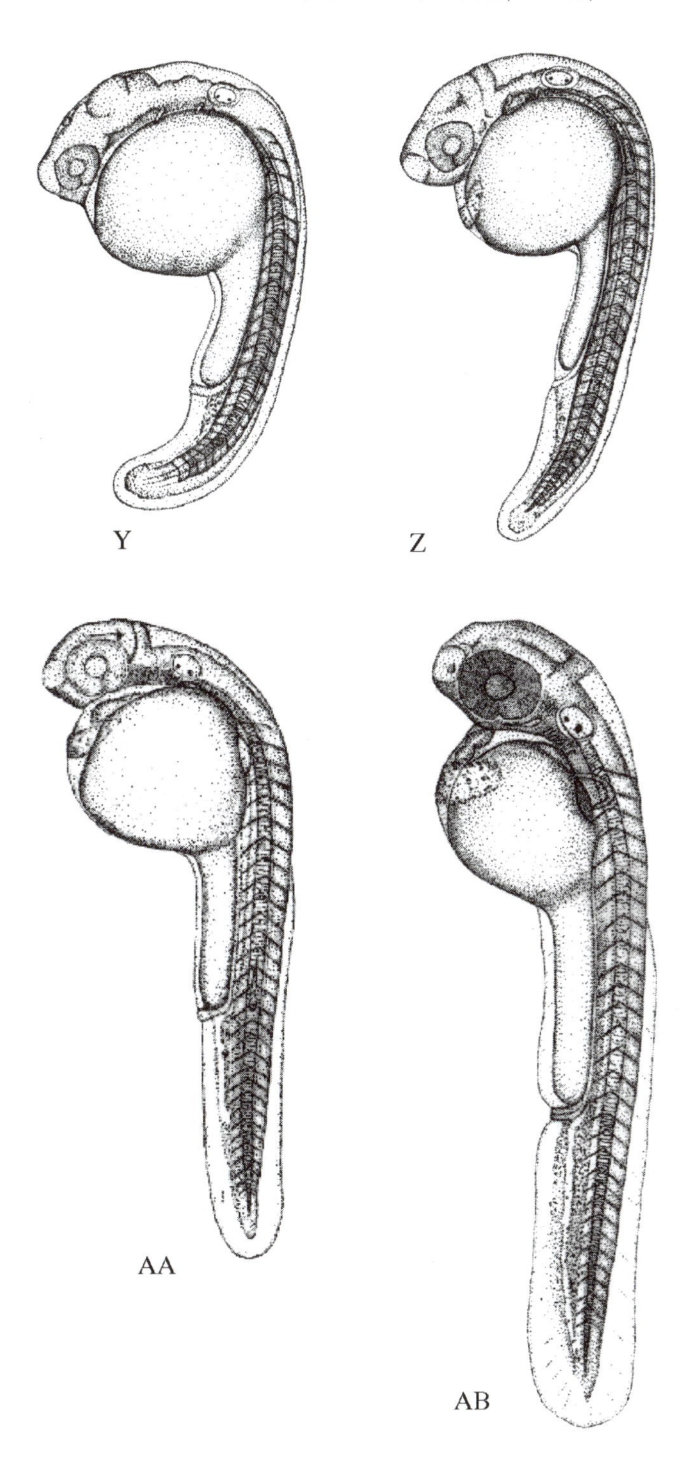

Fig. 1.8 Continued

Table 1.2 Hamburger–Hamilton stages of normal chick development

Stage	Description
1	Prestreak (0–5 h) The embryonic shield may be visible, but the primitive streak has not formed
2	Initial streak (6–7 h) The primitive streak appears as a short conical thickening at the border of the area pellucida
3	Intermediate streak (12–13 h) The primitive streak extends to the center of the area pellucida
4	Definitive streak (18–19 h) The primitive streak is at maximum length, with a primitive groove, primitive pit, and Hensen's node present. The area pellucida is shaped like a pear, and the primitive streak extends 2/3 to 3/4 of its length
5	Head process (19–22 h) The notochord is visible, but the head fold has not formed
6	Head fold (23–25 h) The head fold is visible, but somites have not formed
7	1 somite (23–26 h) One pair of somites is visible, and neural folds are present
8	4 somites (26–29 h) The neural folds are beginning to fuse, and blood islands are present
9	7 somites (29–33 h) The optic vesicles are present, and the heart tubes are beginning to fuse
10	10 somites (33–38 h) Three primary brain vesicles are visible. The optic vesicles are not constricted
11	13 somites (40–45 h) Five neuromeres are visible in the hindbrain. The optic vesicles are constricted at their base, and the heart is bent to the right
12	16 somites (45–49 h) The telencephalon is visible; the auditory pits are deep; the heart is S-shaped. The head fold of the amnion covers the telencephalon
13	19 somites (48–52 h) The head is turning to the right; the telencephalon is enlarged; the amniotic fold covers the head to the hindbrain. Cranial and cervical flexures are present
14	22 somites (50–53 h) The cranial flexure equals about 90°; branchial arches 1 and 2 and grooves 1 and 2 are distinct; the optic vesicles are invaginated as optic cups, and the lens placodes are present
15	(50–55 h) The cranial flexure is more than 90°; branchial arch 3 and groove 3 are distinct; the optic cups are well formed
16	(51–56 h) The wing buds are visible; the tailbud is present; the leg buds are not yet visible
17	(52–64 h) The wing and leg buds are visible; the epiphysis is distinct; the nasal pits are forming; the allantois is not visible
18	(3 days) The leg buds are slightly longer than the wing buds; the amnion is nearly or completely closed; the cervical flexure equals about 90°; the maxillary processes and 4th grooves are indistinct or absent; the allantois is visible
19	(3–3½ days) The maxillary processes are distinct and are as long as the mandibular processes; the allantois is a small pocket but not yet vesicular; the eyes are unpigmented
20	(3–3½ days) The trunk is straight; the allantois is vesicular and about as large as the midbrain; the eye is slightly pigmented
21	(3–3½ days) The limb buds are slightly asymmetrical; their axis is directed caudally. The maxillary processes are longer than the mandible, extending to the middle of the eye. The 4th arch and groove are distinct; the allantois extends to the head; eye pigmentation is distinct

Source: Hamburger and Hamilton (1951).

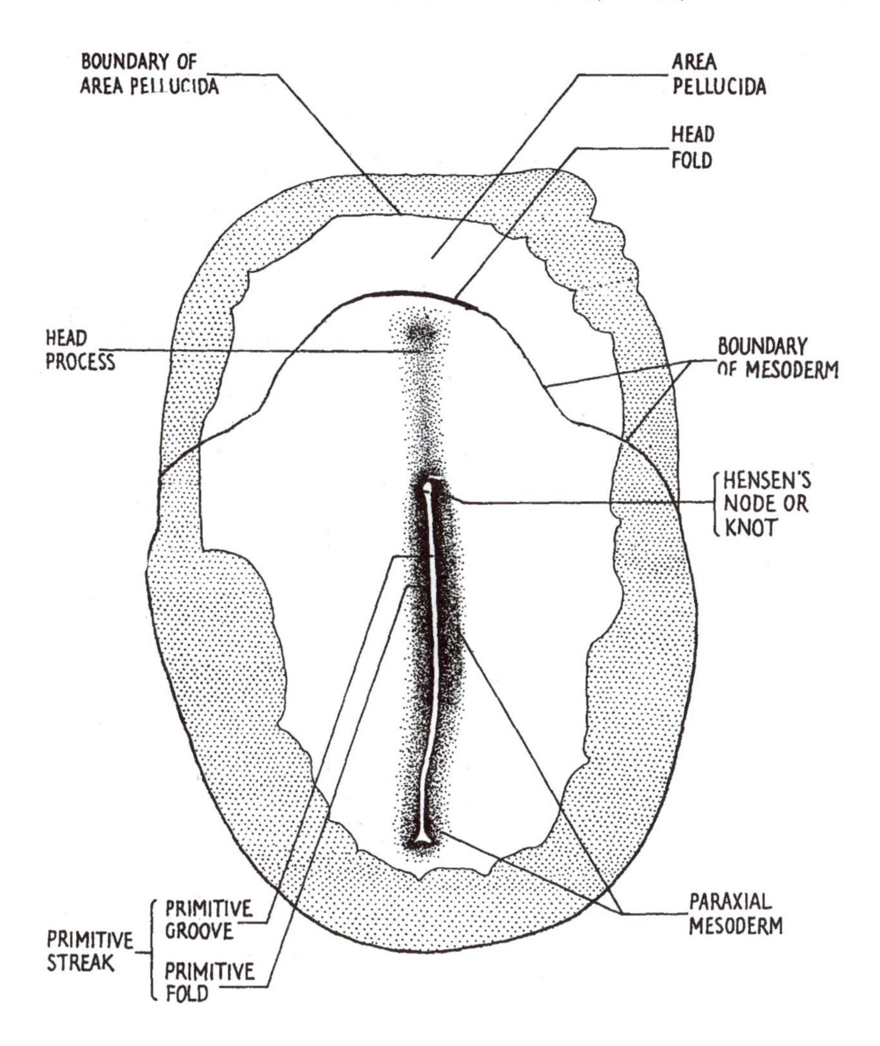

Fig. 1.9 Normal chick development. Stage 5 (19–22 hours). (After W. H. Freeman and B. Bracegirdle, *An atlas of embryology*, Heinemann Press, 1962.)

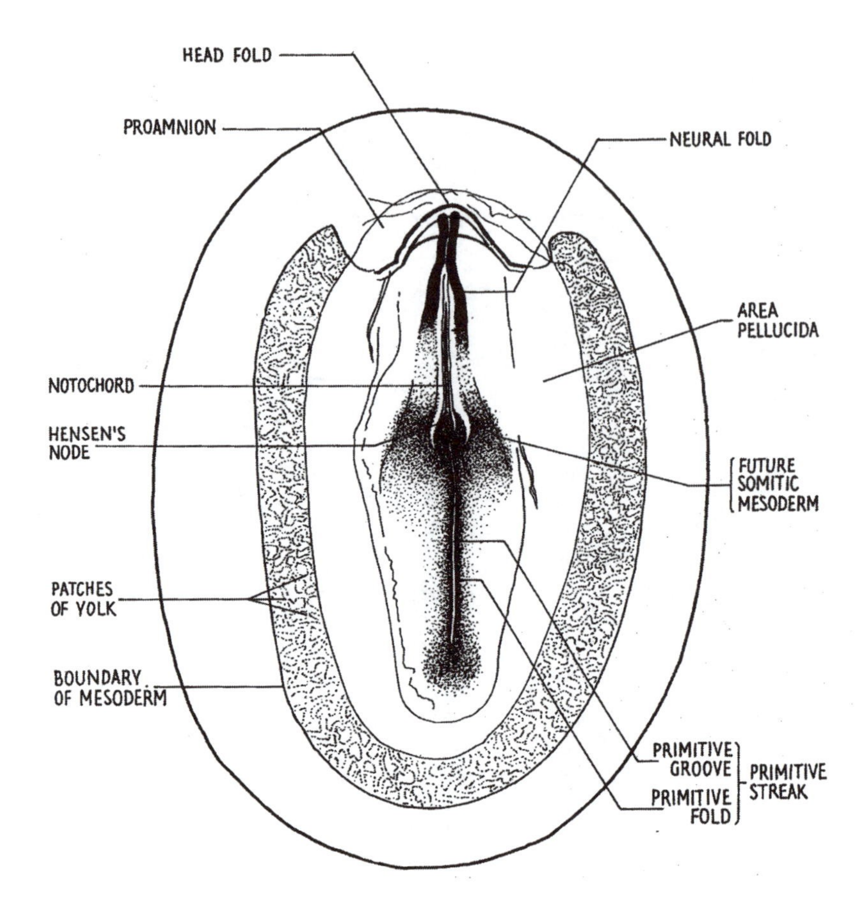

Fig. 1.10 Normal chick development. Stage 6 (23–25 hours). (After W. H. Freeman and B. Bracegirdle, *An atlas of embryology*, Heinemann Press, 1962.)

Plants

The common bush bean has large, rapidly germinating seeds and, therefore, is an excellent tool for observing angiosperm embryonic development. The protective coat on seeds is the aril, an outgrowth of the mature ovule. This coat can easily be removed, after soaking the seed in water for several hours, to reveal the embryonic structures. The cotyledon provides a store of energy until the plant can begin to photosynthesize, and can even serve as a site for photosynthesis until leaves are well formed. The radicle is the root primordium, the hypocotyl is the shoot primordium, and the plumule is the first leaves (Fig. 1.14).

Bean germination protocol

Working on your own:

1. Fold two presoaked bean seeds in damp paper towels, place them in large plastic Petri dishes and keep one near a window and the other in a dark drawer.

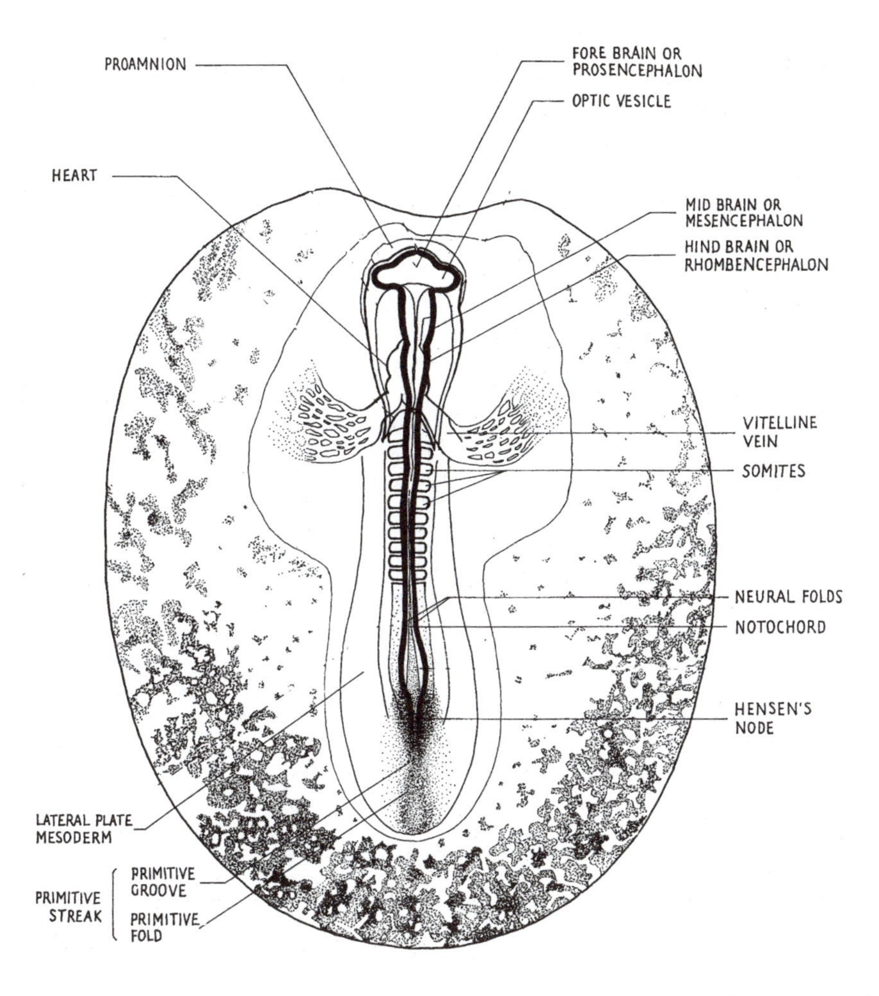

Fig. 1.11 Normal chick development. Stage 10 (33–38 hours). (After W. H. Freeman and B. Bracegirdle, *An atlas of embryology*, Heinemann Press, 1962.)

2. Keep the paper towels moist and record your beans' progress every day.
3. Note how the two treatments compare.

Sporocarps of the water clover *Marsilea* give us the opportunity to watch the production of male and female gametophytes from micro- and megaspores, respectively (Fig. 1.15), from there zygote development, and, if you're lucky, a sporophyte. You will be given a sporocarp (a hard, woody structure), which contains both micro- and megaspores. The small microspores (~0.07 mm) are spherical and develop into male gametophytes. The male gametophyte contains two antheridia, each of which produces 16 sperm. The megaspores, which produce a female gametophyte, are relatively large (0.75 mm) ovals. The female gametophyte includes an archegonium and its egg. After fertilization, the zygote will develop inside the archegonium. The first true sign of zygotic development will be a bright-green sheath that will develop around the nascent first leaf. Within several days, a leaf and root will appear.

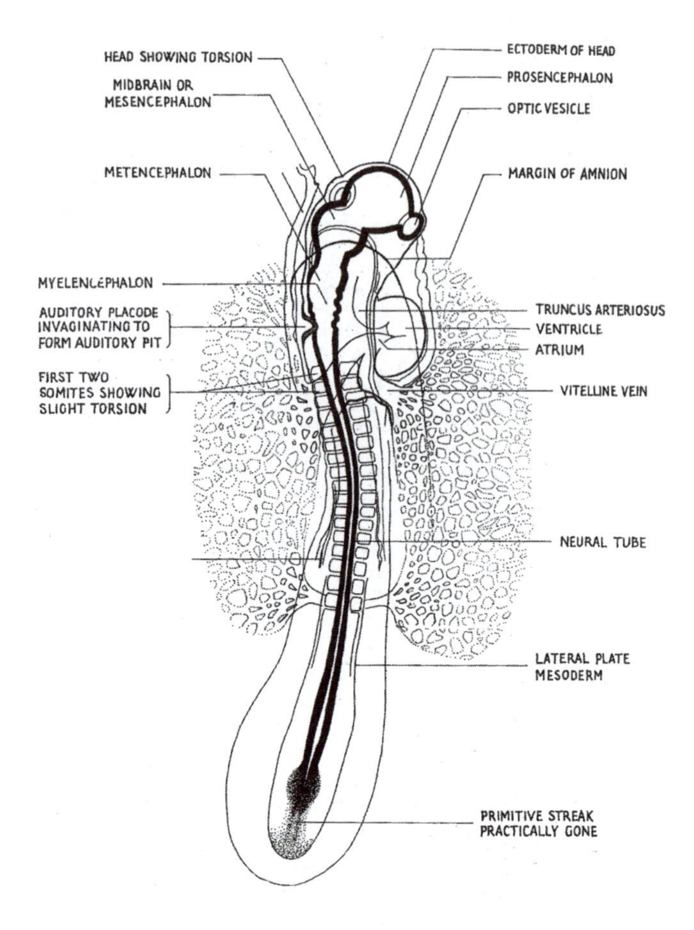

Fig. 1.12 Normal chick development. Stage 12 (45–49 hours). (After Freeman and Bracegirdle 1962.)

Marsilea germination protocol

1. Cut the sporocarp in half with a single-sided razor blade. The sporocarp is very hard, so you will need to exert a fair amount of force to cut through it. If any bits of the sporocarp fly off the table (and they will!), recover the pieces.
2. Immerse the pieces of sporocarp in the Fern medium in your Petri dish.
3. A colloidal substance inside the sporocarp will immediately start absorbing liquid and will expand out of the sporocarp, carrying the spores with it. The spores will be attached to the colloidal substance.

   (a) How much time elapsed between cutting the sporocarp and the emergence of the colloidal material?

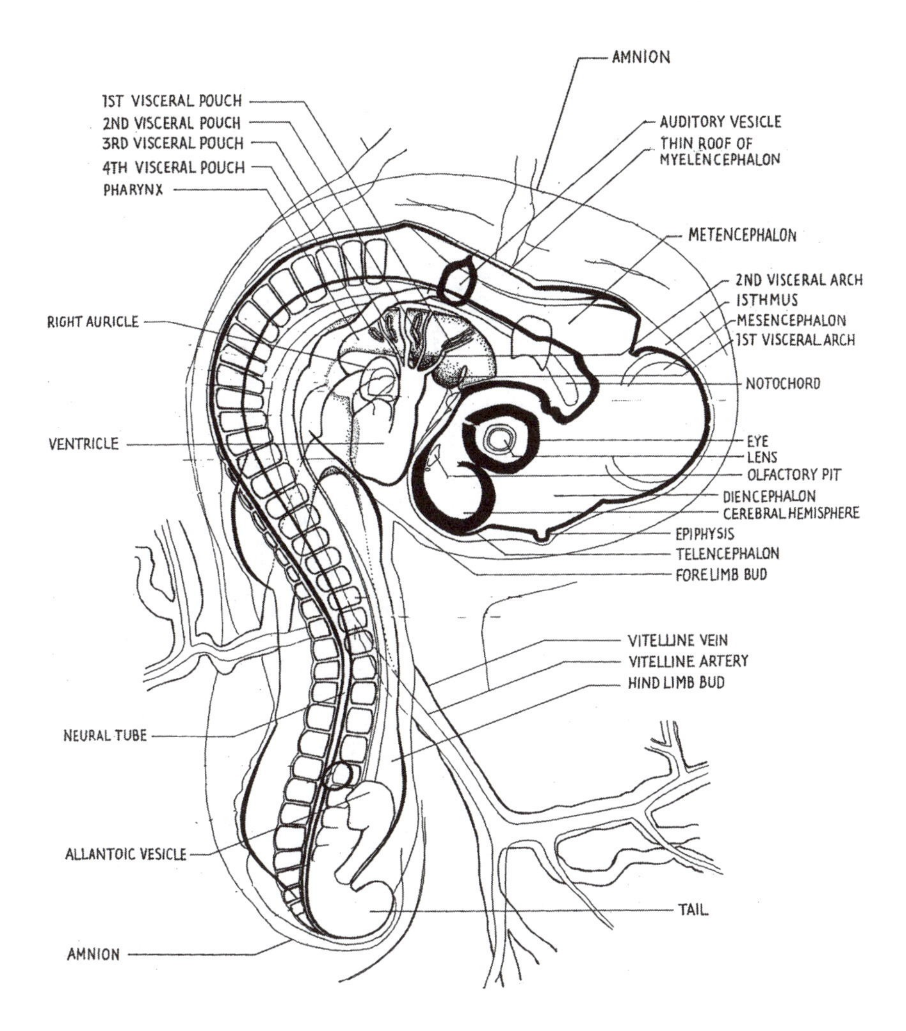

Fig. 1.13 Normal chick development. Stage 18 (3 days). (After W. H. Freeman and B. Bracegirdle, *An atlas of embryology*, Heinemann Press, 1962.)

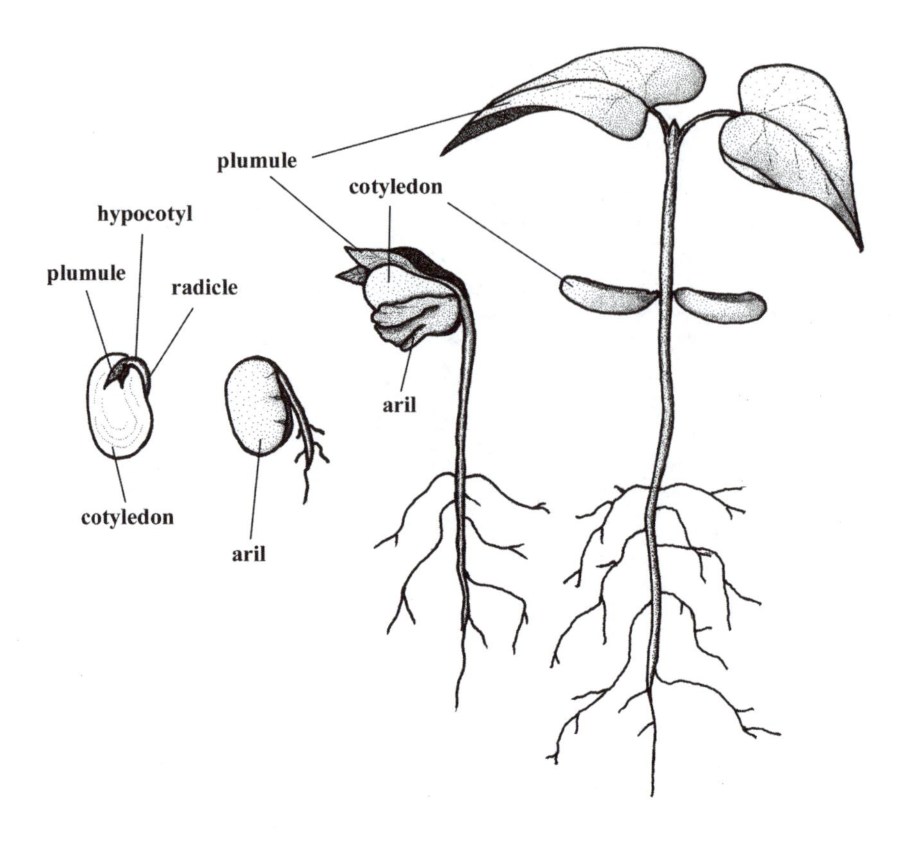

Fig. 1.14 Development of a bean plant from a germinating bean embryo.

(b) How many megaspores and microspores are there in each cluster?

(c) How long do the spores remain clustered before they disperse in the medium?

4. Try to see when swimming sperm are released. Describe any changes you see in the microspores before sperm release.

5. Fertilization should occur within 10 hours of releasing the spores. You should move your spores into a fresh dish within 14 hours of fertilization to prevent bacteria from getting a foothold in your culture.

6. Transfer a good number of female gametophytes to a fresh Petri dish containing Fern medium.

7. Sporophytes will emerge from the female gametophyte within 2–3 days of fertilization. Once your sporophyte develops roots and leaves, you will be able to transplant the organism to a pot of very moist soil and continue to make observations until it becomes a mature adult.

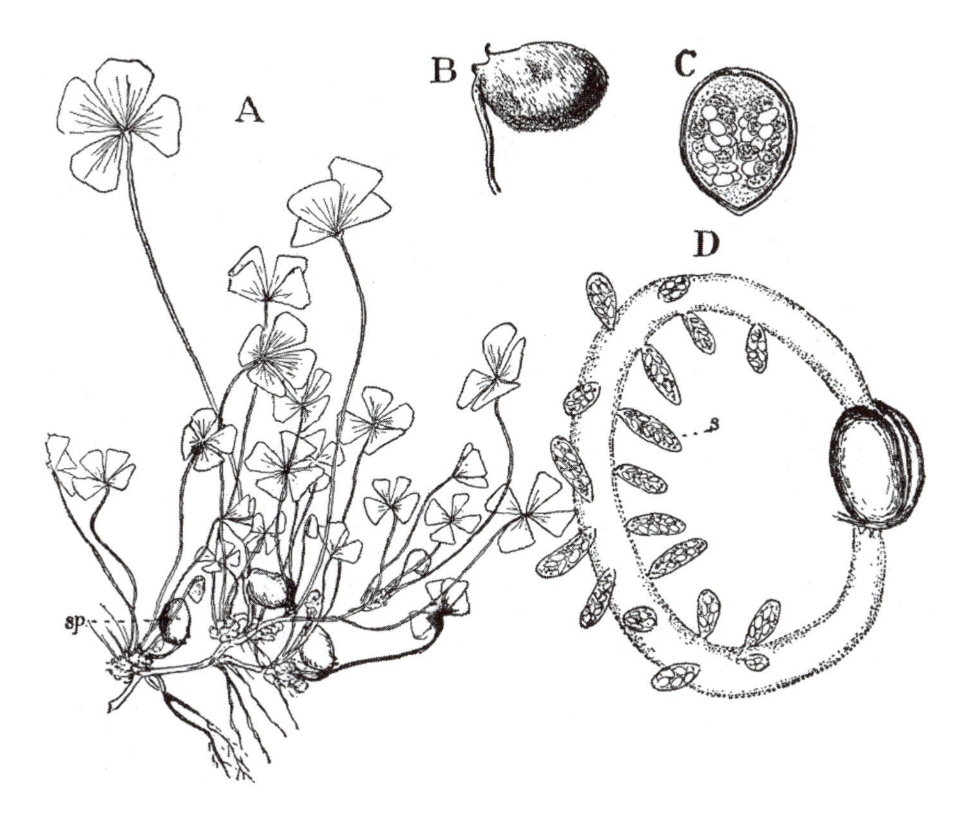

Fig. 1.15 (A) *Marsilea* plant; (B) sporocarp; (C) sectioned sporocarp showing the mega- and microspores inside; (D) germinating spores carried out of the sporocarp by a newly hydrated gelatinous ring (made of a colloid). (From D. H. Campbell, *The structure and development of the mosses and ferns*, 2nd edn, MacMillan Press, 1905.)

Videos/CDs

Vade mecum, *an interactive guide to developmental biology* (CD)—Sinauer Associates, 23 Plumtree Rd, Sunderland, MA 01375.

A dozen eggs, time-lapse microscopy of normal development (ed. Rachel Fink)—produced and distributed by Sinauer Associates for the Society for Developmental Biology.

2 AXIAL PATTERNING

Using retinoic acid to disrupt homeobox gene expression in axolotls

1 week

Homeobox (*Hox*), or homeotic, genes are responsible for patterning the anterior–posterior axes of all animals and plants. Members of this family of genes were first identified in *Drosophila* (Gehring 1998); however, we have since found homeobox genes in all insects, nematodes, cnidarians, platyhelminthes, and vertebrates. These genes are expressed along the anterior–posterior axis in such a way that some may be expressed along the entire length of the axis, while others are expressed for shorter distances. In addition to this spatial pattern of expression, *Hox* genes are also expressed in a temporal pattern, in which anterior genes are expressed before posterior genes. This gene pattern means that the identity of specific zones or segments of the body are based on the expression of a combination of different *Hox* genes (*Hox* clusters).

Hox genes are controlled by a variety of signals, including retinoic acid (RA). Retinoic acid is a vitamin A derivative that is both a necessary endogenous signal (lack of vitamin A during pregnancy is known to cause craniofacial deformities) and a potent mutagen (a related substance, RetinA is used to treat acne, but it carries a warning against using it during pregnancy). Often, interference with homeobox genes results in homeotic transformation. Homeotic transformations are often a small shift in identity; e.g. midbrain becomes hindbrain, anterior head becomes midhead, etc. These transformations may also be quite drastic; e.g. mutations of the Antennapedia complex result in legs growing where antennae should be! *Hox* gene expression also occurs in such a way that posterior genes tend to dominate over genes that are normally expressed anteriorally. This means that when mutations are induced, they have a much greater effect on anterior structures (Gehring 1998) (Fig. 2.1).

In today's experiment you will be exposing gastrulating and neurulating axolotl embryos to several different concentrations of all-*trans*-retinoic acid and looking for induced deformities during the following week. The most likely changes you will see will be a loss of anterior head structures.

WARNINGS

- ***!*** *Retinoic acid is a teratogen and therefore must be handled with all due precautions. It must not be poured down the drain! Dispose of it properly.*
- ***!*** *DMSO (dimethylsulfoxide) is being used to help move retinoic acid across the cell membranes of axolotls and will also move chemicals across your own membranes.*

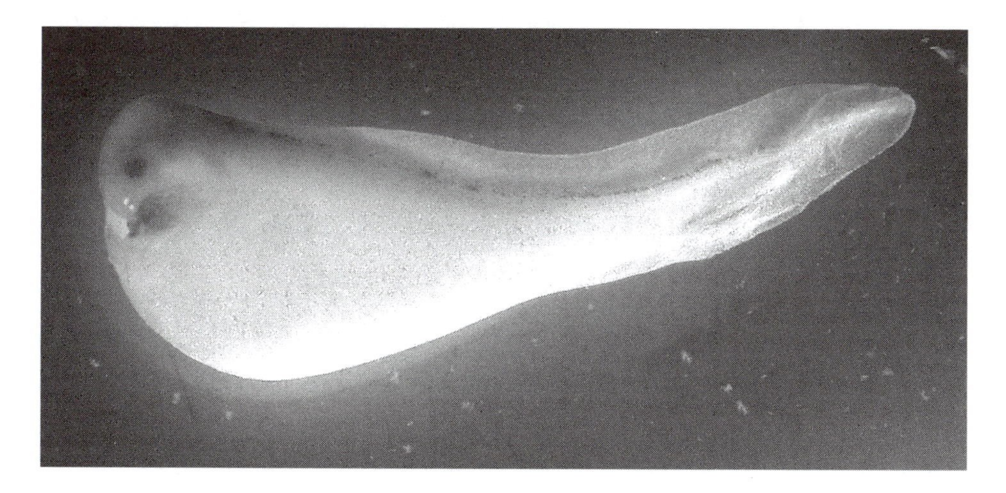

Fig. 2.1 Axolotl embryo exposed to 10^{-7} mol/l retinoic acid. Notice the tiny head and reduced gills (directly under the eye). Pigmentation is also reduced.

Use gloves when handling DMSO and keep the bottle lid on as much as possible (fumes). Waste should not be poured down the drain.

! *Glutaraldehyde is corrosive and can cause skin and respiratory irritation. Wear gloves and use a fume hood. Waste should not be poured down the drain.*

TROUBLESHOOTING

Controls died

- Embryo clutch was of poor quality.
- Rearing solution was contaminated (make fresh RS).
- Antibiotics were left out of the rearing solution (make fresh RS).
- Glassware was contaminated (clean and sterilize glassware).
- Distilled water source was not clean.
- Poor sterile technique employed (review general embryo protocols).

Controls deformed

- Embryo clutch was of poor quality.
- Embryos were damaged during the dejellying procedure (practice dejellying again until you can easily dejelly embryos without damaging them).

Experimental animals not deformed

- Retinoic acid inactivated by being stored above 4 °C (this will happen if you make your RA solution more than an hour or two ahead of the lab session!). (Obtain fresh retinoic acid.)
- Retinoic acid dilutions were improperly made (recalculate dilutions of retinoic acid in RS).

Materials

2 pairs of fine forceps

Sterile Petri dishes

Sterile glass or plastic Pasteur pipettes

Sterile graduated glass pipettes

Sterile glass vials (20 ml)

Micropipettor

Shade for Petri dishes (black-paper tent construction, dark cloth, or put in a box or drawer)

Waste containers for RA-, DMSO-, and glutaraldehyde-contaminated solutions

Solutions

All-*trans* retinoic acid

Rearing solution

DMSO (dimethylsulfoxide)

0.01% formalin fungicide solution

70% ethanol

MS-222

4% glutaraldehyde in PBS

Protocol

1. Work in groups of two or three.
2. Sterilize forceps in 70% ethanol for 10 minutes.
3. Sterilize 20 embryos (still in jelly coat) in the formalin solution for 10 minutes.
4. Label four small Petri dishes with test RA concentrations and control.
5. Using a cut-down plastic pipette, transfer embryos to a large shallow Petri dish filled with RS. Be careful to transfer as little of the formalin as possible.
6. Dejelly the embryos and transfer them into a new Petri dish filled with fresh RS.
7. Prepare retinoic acid solutions:

   (a) Obtain one aliquot (20 µl) of retinoic acid from the freezer (10^{-2} mol/l dilution).

(b) To make an ~10^{-5} mol/l dilution, take 10 μl of the stock and add to 10 μl DMSO and 10 ml RS.

(c) To make an ~10^{-6} mol/l dilution, take 1 ml of the 10^{-5} mol/l solution and add 10 μl DMSO and 9 ml RS.

(d) To make an ~10^{-7} mol/l dilution, take 1 ml of the 10^{-6} mol/l solution and add 10 μl DMSO and 9 ml RS.

(e) Your control will be 10 ml RS with 10 μl DMSO.

8. Pour each RA solution (10^{-5} mol/l, 10^{-6} mol/l, 10^{-7} mol/l) and the control into separate Petri dishes and add at least three embryos to each dish.

9. Cover the dishes to shield them from light (RA is light-sensitive) and incubate for 1 hour at room temperature (i.e. between 21 °C and 26 °C).

10. Rinse embryos 3 times by transferring them to a dish of fresh RS containing 1% DMSO; each rinse should last for 10 minutes. The final rinse should be in pure RS (i.e. no DMSO).

11. Rear embryos for 2–2½ weeks, changing the RS every 2–3 days, until they reach hatching stage. You should sketch the embryos each time you change their RS.

12. Sacrifice the larvae in MS-222 (this is also known as tricaine, methansulfonate or 3-aminobenzoic acid ethyl ester, and is used as an anesthetic in fish), preserve in 4% glutaraldehyde (in PBS), and store in 70% ethanol for further analysis.

13. Measure relative body length, head size, eye size, and degree of edema of a representative embryo at each RA concentration (and the control).

QUESTIONS

1. At what concentration of RA do you first see an effect?
2. Describe the RA-induced effects in as much detail as possible. Compare and contrast early and late embryo effects.
3. Are all embryonic stages equally sensitive to RA? Explain your answer.

3 PLANT CELL TOTIPOTENCY

Growing a carrot plant from adult cells

semester-long

Plant reproduction does not rely entirely on gametes; individual adult somatic cells retain the ability to dedifferentiate and develop into new plants! In most animal embryos, on the other hand, this sort of developmental flexibility doesn't last much longer than the early gastrula stage. This regenerative ability is one of the most striking differences between animals and plants. We have been culturing plant tissue successfully since the 1930s, when plant growth hormones like auxin were first discovered. However, it wasn't until 1958 that Steward *et al.* were able to culture an embryo from an adult carrot cell. When plant somatic cells are removed from a mature plant (explant) and cultured in a special medium that contains nutrients, plant growth regulators (like auxins, which induce proliferation while inhibiting differentiation), vitamins, and sugars, they will be stimulated to form a callus (mass of undifferentiated cells) (Fig. 3.1). The callus can then be further stimulated (by a second medium low in auxin but containing the cytokinin kinetin, which promotes cell division and embryogenesis) to differentiate into root and shoot cells.

The purpose of this lab session is to explore dedifferentiation of adult plant cells to form a normal embryo. You will explore animal cell dedifferentiation in a later laboratory session.

TROUBLESHOOTING

- Contamination is the major problem in this experiment. Sterile techniques must be strictly maintained at all stages of the experiment.

Materials

Carrot Callus Initiation Medium in sterile containers

Carrot Shoot Development Medium in sterile containers

Large, sterile plastic Petri dishes

Cork borer

Scalpel

70% ethanol

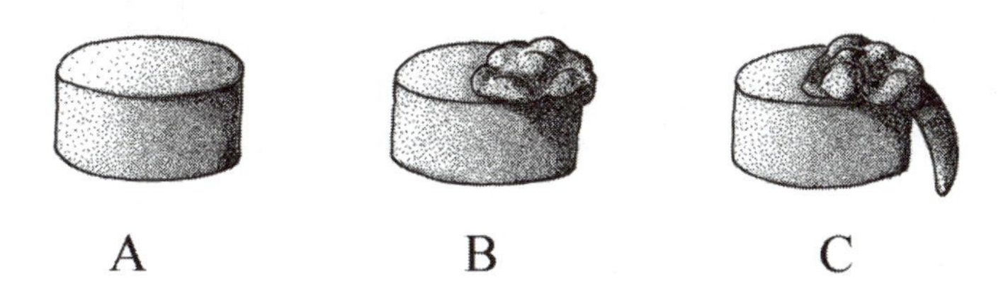

Fig. 3.1 (A) Carrot explant (disc) cells proliferate to form a callus (B) and then a rootlet (C).

Sterile forceps (soak in 70% ethanol)

Sterile glass rods

Carrot (>5 cm diameter), washed with soap

Sterile workstation (laminar flow hood)

Protocol

1. If you are using a hood with a germicidal (UV) light, turn it on for 15 minutes before beginning your experiment and turn it off before using the hood; UV light is only used to disinfect the hood. **Do not** work with the UV light on, it can damage your eyes.
2. Spray-down the workstation with 70% ethanol.
3. Break the ends of the carrot off, being careful not to touch the freshly exposed carrot tissue. You will need 5–7 cm of carrot.
4. Position the larger end of the carrot in a Petri dish and push the cork borer into the smaller end of the carrot.
5. Use the glass rod to push the carrot tissue out of the borer and into a fresh Petri dish.
6. Cut off the ends of the core and discard. Cut the rest of the core into discs (2–4 mm).
7. Place the discs onto the Carrot Callus Initiation Medium and place the dish in indirect sunlight or under fluorescent lights.
8. Check your discs every few days for contamination (remove any contaminated discs).
9. Sketch your discs every week. After 2–3 weeks you should start seeing a callus form on the discs.
10. After the calluses have grown (4–6 weeks after placing them in the callus medium), transplant the discs to Carrot Shoot Development Medium using the same sterile techniques.

11. Place the calluses in indirect light for 4 weeks.

12. Look for changes in the appearance of your calluses, especially clusters of cells that appear to be growing away from the original callus.

13. After 8–10 weeks, the new roots and shoots should be large enough to plant.

QUESTIONS

1. What percentage of your carrot discs formed a callus?
2. What percentage of your discs formed rootlets?
3. What types of contaminants gained access to your carrot discs? (You may want to take samples and identify them in a microbiology lab.)

FERTILIZATION

Sea urchin syngamy and development

4

2–3 days

Having read about sea urchin development in your textbook, today is your chance to watch the fusion of sea urchin eggs and sperm and the development of the embryo into a pluteus or feeding larva. Since urchins are spawners (external fertilization) and gametes have a very limited lifespan, it is important that males and females release gametes at the same time. This release can be triggered by photoperiod, water temperature changes, and stress. As an added encouragement to synchronous spawning, newly released gametes exude chemicals that trigger conspecific urchins to spawn as well. The mechanism of spawning involves a contraction of the gonad wall to force the eggs or sperm out through the five gonopores on the aboral surface (opposite the mouth) (Fig. 4.1) and into the water column. We will not be relying on natural stimulants to induce spawning, and since there is no external sexual dimorphism in sea urchins, we must rely on a general stimulus that will induce both males and females to spawn. We can stimulate sea urchins to spawn by injecting the perioral tissue with an isotonic KCl (potassium chloride) solution or exposing them to a mild (6 V) electric current.

Sea urchin eggs are covered with a thin (~35 μm) transparent jelly coating that is not visible under a compound microscope unless stained (a dilute India ink solution will allow you to visualize it). The eggs actually draw the sperm towards them by exuding a sperm attractor signal! When a sperm cell makes contact with the egg, it sets in motion a series of reactions designed to prevent polyspermy. The fast block of polyspermy is a transient depolarization across the cell membrane that prevents other sperm from fusing with the egg cell membrane. This current change lasts just long enough for the slow (permanent) block changes to begin. Slow block starts with a release of calcium, followed by the release of cortical granules, expansion of the vitelline membrane away from the egg (this is visible under the microscope! So keep your eyes open!), and, finally, the fertilization (vitelline membrane) and hyaline (egg membrane and cortical granule materials) layers are formed.

TROUBLESHOOTING

- There is a good chance that when you receive your urchins in the mail, they will begin to spawn spontaneously, setting off a spawning chain reaction. If possible, you should plan your order so that the urchins arrive on the day of your lab period. The urchins should be transferred to a tank of fresh seawater as soon as possible, although you

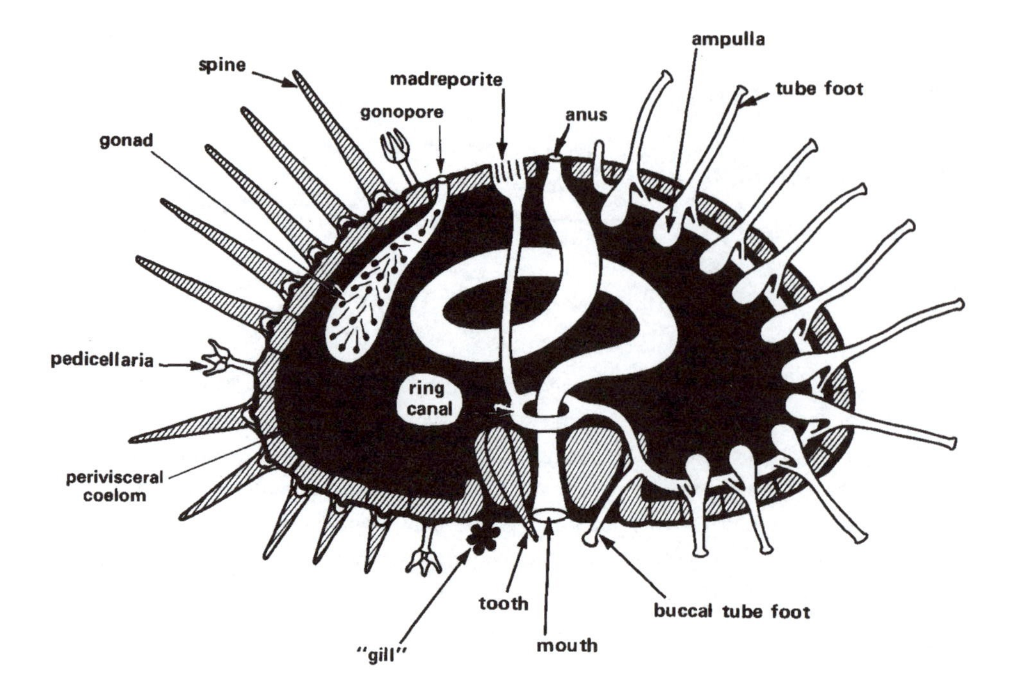

Fig. 4.1 Internal anatomy of a sea urchin. (Reprinted from V. Pearse, J. Pearse, M. Buchsbaum, and R. Buchsbaum, *Living invertebrates*, Boxwood Press 1987, with permission of the authors.)

may still find it necessary to siphon gametes off madly spawning urchins and save them for your lab session (they will remain viable for several hours if left undiluted or rinsed).

- If, after injections of KCl, the urchins do not begin to shed gametes, you may have to sacrifice the animals to extract gametes by dissection (cut them open with scissors).
- If embryos fail to develop past the blastula stage, check for proliferating microorganisms. This is not a sterile preparation, so it isn't uncommon for *Paramecium* spp. or bacteria to be present in the culture.

Materials

Plastic pipettes

Depression slides

Compound microscope

Fingerbowls or beakers

0.5 mol/l KCl

Syringes

Sea urchins

Small Petri dishes

Instant Ocean (artificial seawater mix)

5-gallon (~19 liters) aquarium

Dissecting scissors

Procedure

1. Inject 0.5 ml KCl into the soft tissue surrounding the mouth on the underside of the urchin. If you have a male, you will begin to see opaque white sperm being released from the top of the urchin. If you have a female, reddish-brown eggs will be shed from the top of the urchin. It may take several (5) minutes before shedding begins. If nothing happens, try a second injection. If there is still no response, you may need to remove the gametes by dissection.

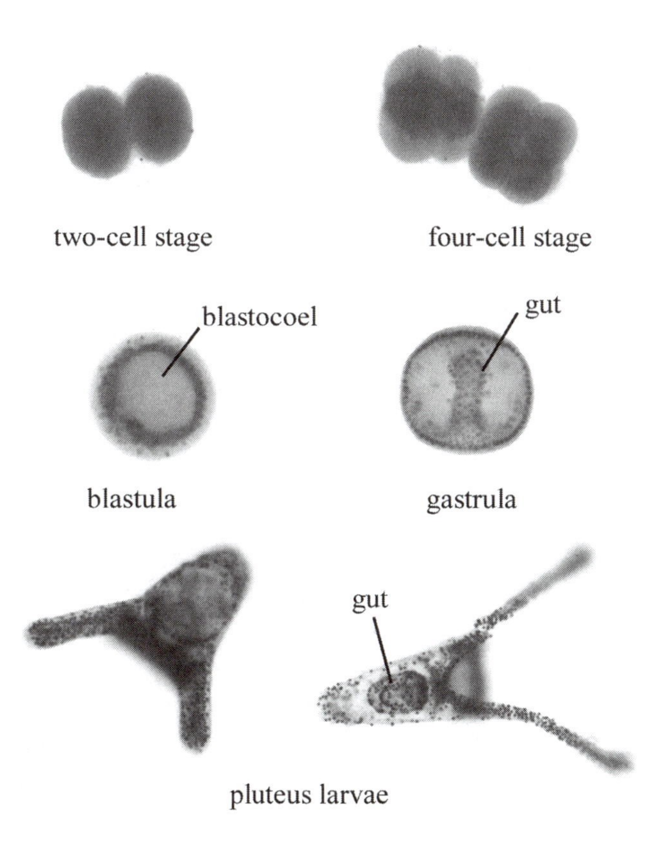

Fig. 4.2 Photomicrographs illustrating distinctive stages of sea urchin development.

2. Position each urchin upside down over a small fingerbowl containing a small amount of seawater to collect the gametes. It may take up to 15 minutes before the urchin finishes shedding gametes.

3. Rinse the eggs in fresh seawater so a new egg 'scent trail' will be released and subsequently followed by the sperm.

4. Take a single drop of sperm and dilute it in 5 ml of seawater.

5. Add a drop of egg solution to a depression slide and, while watching, add a drop of diluted sperm and watch the show (vitelline membrane expansion)!

6. Draw, in as much detail as you can, how the egg (or is it now an embryo?) looks.

7. Prepare an urchin development chamber in a small Petri dish for the rest of your eggs. For every milliliter of egg solution, add a few drops of the dilute sperm solution.

8. Allow the eggs and sperm to mix for about 20 minutes; swirling gently every few minutes.

9. Observe a sample every 30 minutes until you see a two-cell stage embryo (<2 h) (Fig. 4.2).

10. During each of the next 3 days, you (or one of your lab partners) should take samples from the development chamber and draw what you see.

QUESTIONS

1. What percentage of eggs were fertilized and began cleavage?
2. What percentage of dividing cells formed blastulae? prism larvae?
3. What percentage of cells died? Do you think this level of mortality could be normal? Explain.

EARLY PLANT DEVELOPMENT

5

Pollen tube formation

1 day

Pollen, which many people may consider to be the plant's version of a sperm, is actually a gametophyte (which **generates** gametes like sperm), contains two nuclei instead of sperm's single nucleus, and will eventually form two sperm cells. Pollen lands, or is deposited, on the stigma and is stimulated, by secretions of the stigma and style (Fig. 5.1), to germinate (begin the formation of a pollen tube) (Figs 5.2 and 5.3). These factors are species-specific and, in some cases, individual-specific, to prevent self-fertilization. Pollen tube growth can be blocked by incompatibility barriers on the stigma or style. So, how does a pollen tube form? Pollen cell walls are well armored with a material called sporopollenin, yet have several germination pores or grooves that are essentially weak points through which the pollen tube may grow. Pollen tubes grow as large intercellular vacuoles that expand, pushing the pollen's cytoplasm and sperm cell nuclei into the tip of the pollen tube. Amazingly enough, the pollen tube, which may be several hundred micrometers long, is mostly vacuole, separated from the cell material by callose plugs. The pollen tube moves easily through the stigma and style, absorbing nutrients as it goes and guided by biochemical gradients. The pollen tube will eventually enter the ovule through the micropyle and fertilization will occur. You will learn more about the anatomy of the ovule and zygote in a later chapter.

Today you will measure and compare normal rates of germination and pollen tube elongation in two species of angiosperms. Germination will be stimulated by normal (plain) pollen tube medium, as well as being exposed to media supplemented with actinomycin D (inhibits transcription), cycloheximide (inhibits translation), and cytochalasin B (inhibits microtubule assembly).

TROUBLESHOOTING

Finding pollen is not always easy. Fresh-looking flowers may have no pollen left on them, so be sure to check the anthers under the dissecting microscope before making your pollen suspension. Also, the anthers must be mature. For example, *Melilotus* grows in a spike; those flowers on the top are too young and those at the bottom are too old.

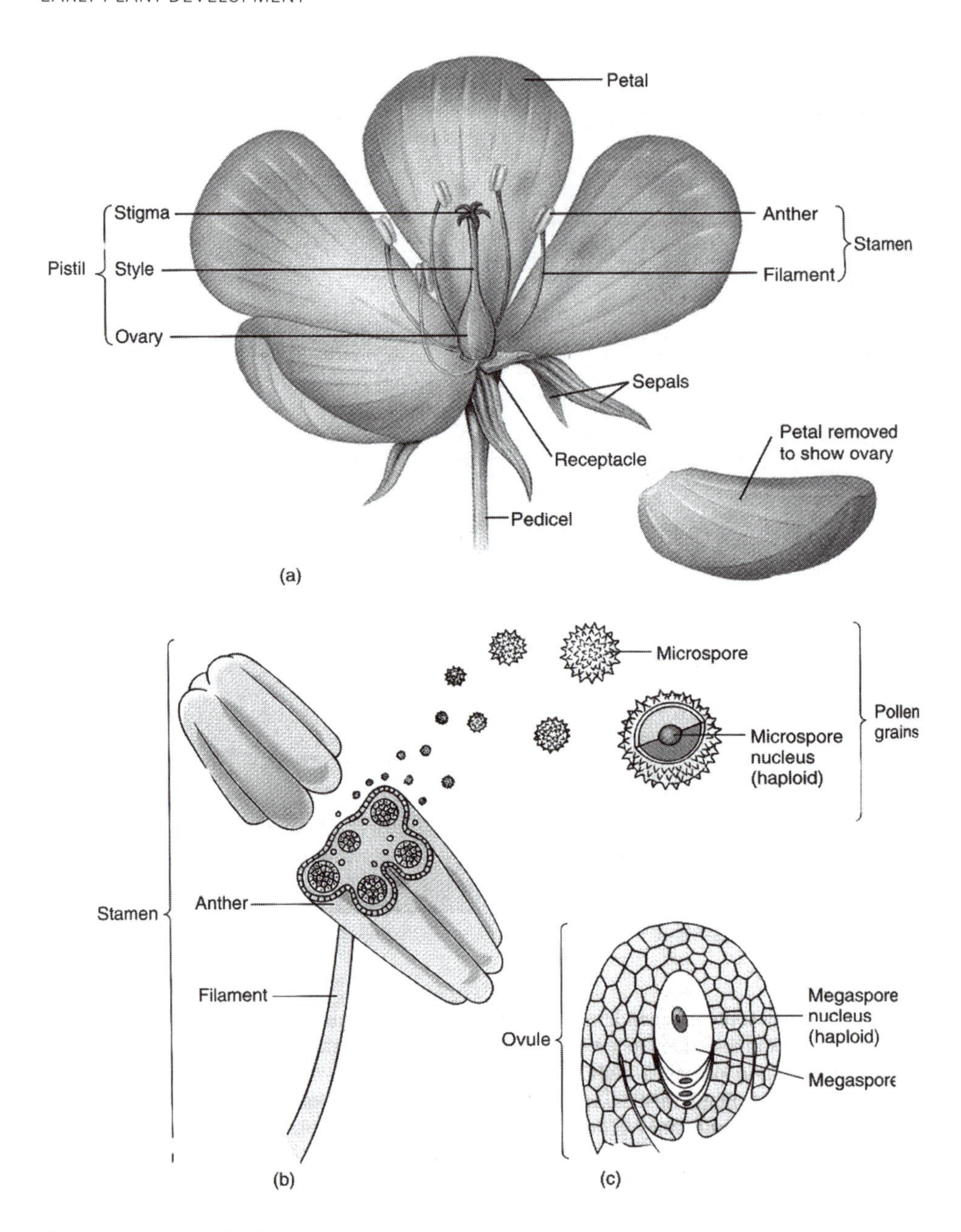

Fig. 5.1 Anatomy of a flower and its reproductive organs. (Reprinted from Mauseth 1998 with permission.)

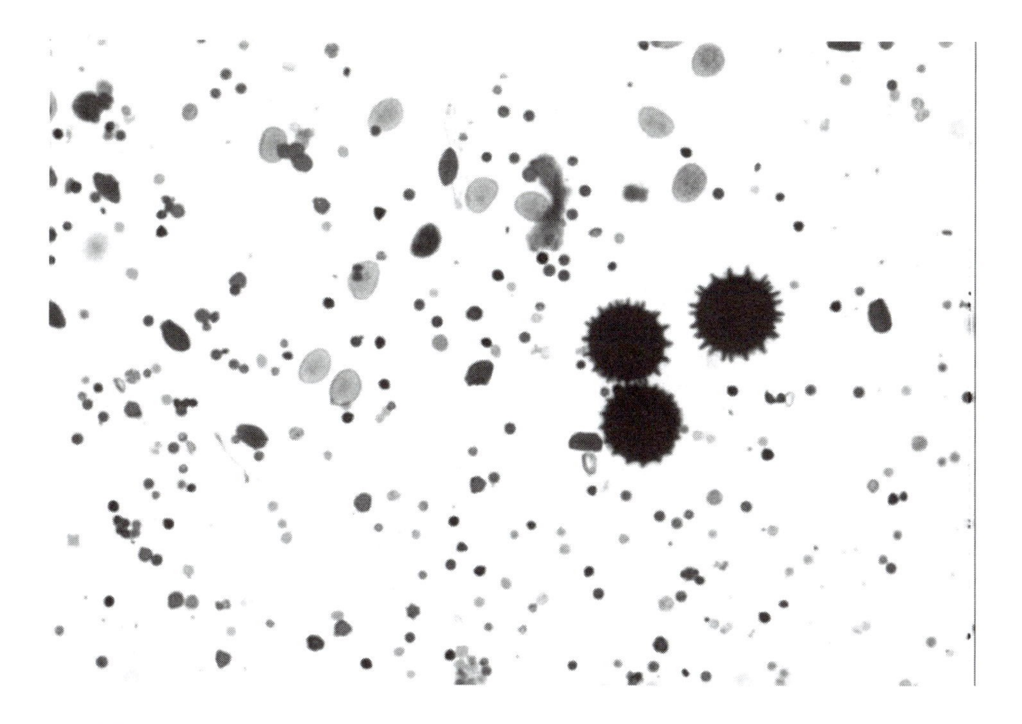

Fig. 5.2 Assorted pollen grains. Note the difference in size, shape, and staining.

Materials

Mature flowers:

- *Melilotus officinalis* (yellow clover)
- *Melilotus alba* (white clover)
- *Gibasis geniculata* (bridal veil)
- *Tradescantia* (spiderwort)
- Sweet pea

Petri dishes (35 × 10 mm)

Plain pollen tube medium

Pollen tube medium supplemented with actinomycin D:

- stock: 100 μl DMSO + 1 mg actinomycin D
- solution: 30 μl stock + 10 ml Plain pollen tube medium

Pollen tube medium supplemented with cycloheximide:

- stock: 1 ml DMSO + 1 mg cycloheximide
- solution: 200 μl stock + 10 ml Plain pollen tube medium

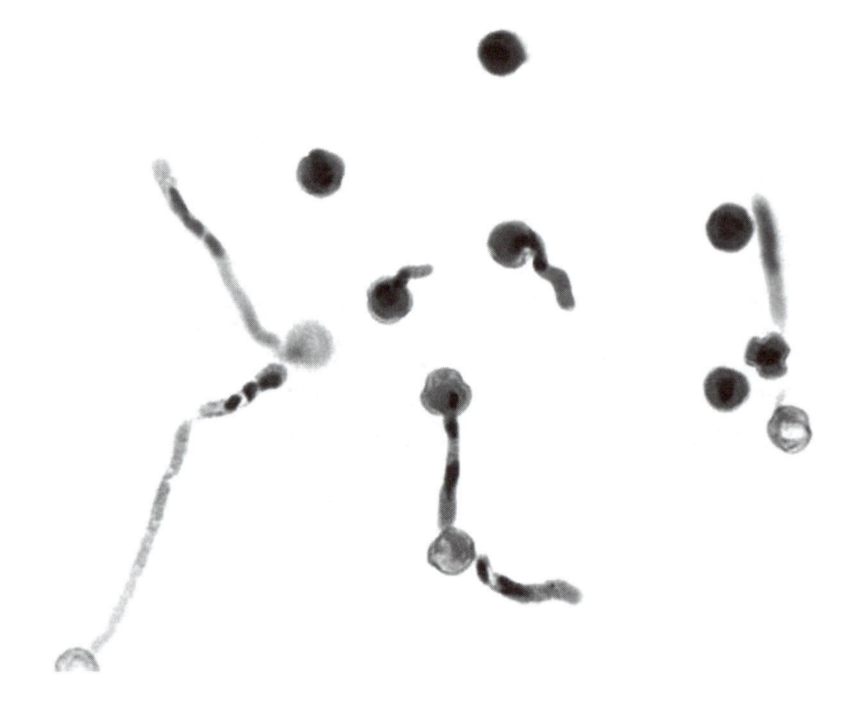

Fig. 5.3 Pollen tubes forming in *Lilium* pollen grains.

Pollen tube medium supplemented with cytochalasin B:

- stock: 100 μl DMSO + 0.5 mg cytochalasin B
- solution: 40 μl stock + 10 ml Plain pollen tube medium

Pipettes

Forceps

Compound microscope with ocular micrometers (if possible)

Dissecting microscopes

Slides and coverslips

Counting device

Procedure (after Scott 1995)

1. Obtain pollen from a flower. You must examine the anthers under a dissecting microscope for the presence of pollen. Avoid using old or withered flowers.
2. Add pollen to 2 ml of a pollen tube medium in your Petri dish—choose Plain pollen tube medium or one supplemented with actinomycin D (blocks

transcription), cycloheximide (blocks translation), or cytochalasin B (prevents actin microfilament elongation):

- for bridal veil or spiderwort, you can dip the flowers and/or anthers in the medium and compress the anther against the bottom of the Petri dish.
- for *Melilotus*, you will need to force the petals open by compressing the base of the flower to expose the anthers.
- for sweet pea, dissect the anthers out and add to the medium.

3. Mix pollen into the medium and check under the compound microscope to be sure that enough pollen is present.

4. At 5-minute intervals, make a wet mount of your pollen suspension in a depression slide and sketch a few representative pollen grains.

5. Assess germination by counting how many of the first 50–100 pollen grains you see under the microscope have germinated (begun to form a pollen tube).

6. Measure the pollen-tube length of at least 10 germinated pollen grains.

7. Continue measurements for 2 hours or until you can no longer accurately measure the pollen tubes.

8. Graph your data.

QUESTIONS

1. How do the three treatments affect pollen-tube growth? Consider the number of tubes formed and the length of tubes.
2. Which treatment did you expect to have the biggest impact on tube formation? Why?

6 MORPHOGENESIS

Creating fate maps of albino axolotl embryos using a vital dye

1 week

Gastrulation is the first, and arguably most dramatic, morphogenetic event during vertebrate development. The embryo changes from a simple sphere of cells with uncomplicated symmetry, heavy vegetal cells (presumptive endoderm) located ventrally, lighter animal cells (presumptive ectoderm) dorsally, and an equatorial band of presumptive mesoderm. By the time gastrulation is complete, the embryo is bilaterally symmetrical (although still spherical), has three layers of tissue in its walls, and has a primitive gut (archenteron) instead of the blastocoel.

Three basic processes are involved:

1. **Involution**—Surface tissues begin to move inside the embryo when bottle cells (presumptive mesoderm) constrict their apical pole. This new indentation will become the blastopore lip; the point across which all entering tissue must pass.
2. **Convergent extension**—Prior to gastrulation, the presumptive mesoderm that will give rise to notochord is located along the equator of the blastula. As the tissue is drawn from two directions into the blastopore, those two lines of cells intercalate and thereby elongate the resulting tissue band.
3. **Epiboly**—In spite of having a large number of cells moving inside the embryo, overall size does not change. This is because the cells of the animal pole go through radial intercalation, during which one ectodermal cell layer merges with the layer above (or below) thereby thinning the ectoderm and expanding its surface area.

The purpose of this lab session is to illustrate where a specific patch of cells moves on the surface of the embryo as the animal develops, and what kind of tissue that specific patch will become (Fig. 6.1).

TROUBLESHOOTING

- It's usually pretty obvious what has gone wrong in this experiment after the fact. There are, however, a few tips that will help prevent the problems.
- Be wary of the sharp edges of your clay depressions; they can and will cut your embryo. Edges of all depressions should be smoothed prior to adding the embryo.

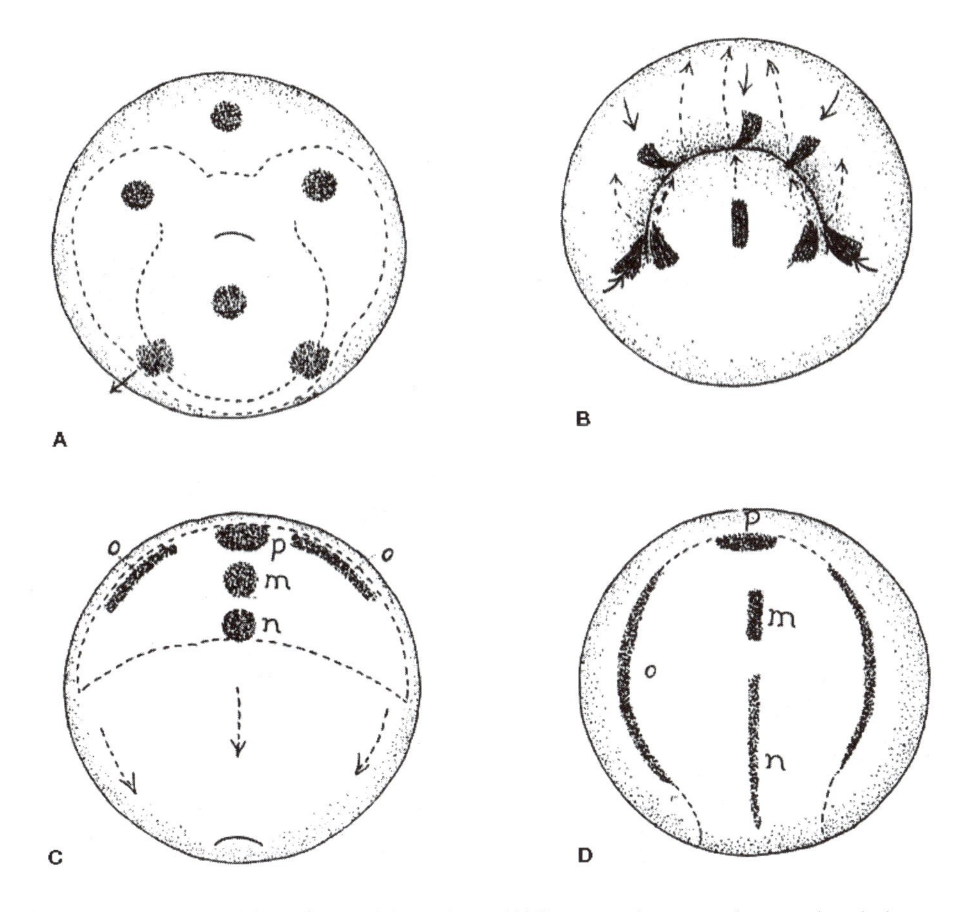

Fig. 6.1 Vital dye staining of a urodele embryo. (A) Presumptive mesoderm and endoderm have been marked with a vital dye. As the embryo gastrulates (B) we see the tissue moving towards the blastopore. A late gastrula embryo (C) has been marked on the presumptive neural plate. The neural plate is clearly delineated in the neural plate stage embryo (D). (Reprinted from V. Hamburger, *A manual of experimental embryology*, The University of Chicago Press, 1960, with permission of The University of Chicago Press.)

- The embryo should only be gently compressed to hold it in the depression. If you overcompress, the embryo will be permanently damaged. Also, do not cover the embryo with the modeling clay; the embryo will not be recoverable.
- If you use large agar chips, you will dye such large areas of the embryo that it will be difficult, if not impossible, to interpret.

Materials

Albino axolotl blastulae

Rearing solution

70% ethanol in a spray bottle

2 pairs of fine forceps

Cut-down plastic transfer pipettes

0.01% formalin fungicide solution

Glass coverslips

Small, sterile Petri dishes

Large, dissecting Petri dish

Petri dish lined with non-toxic (children's) modeling clay

Sterile glass rods

Neutral Red-stained agar chips

Colored pencils

Vital dye protocol

1. Select 8–10 healthy embryos (this number will allow for the usual destruction of embryos that comes with dejellying them) and place them in fungicide for 10 minutes.
2. Sterilize the clay-lined Petri dish with alcohol (use a spray bottle).
3. Transfer embryos into the Rearing solution (RS) in the dissecting dish.
4. Remove the jelly coat, while leaving the vitelline membrane intact (see Chapter 1 for details).
5. Using a sterilized glass rod, make several depressions in the modeling clay that are slightly larger than your embryos. Make sure to smooth the edges of the depressions so the embryo isn't damaged by the sharp edges of the modeling clay. Fill the dish with RS.
6. Cut several very small (≤1 mm) pieces of the dyed agar and set aside.
7. Position an embryo in one of the depressions (Fig. 6.2) and orient it so you can see the area you want to stain. Since the vegetal pole is heavier than the animal pole, you may have some difficulties positioning your embryo. By gently compressing the clay next to the depression holding your embryo, you will be able to fix the embryo in the desired position without killing it.
8. Place the dyed agar in position on the embryo and gently lay a glass bridge across the agar and embryo to hold it in place. You should stain three embryos and keep three as controls.

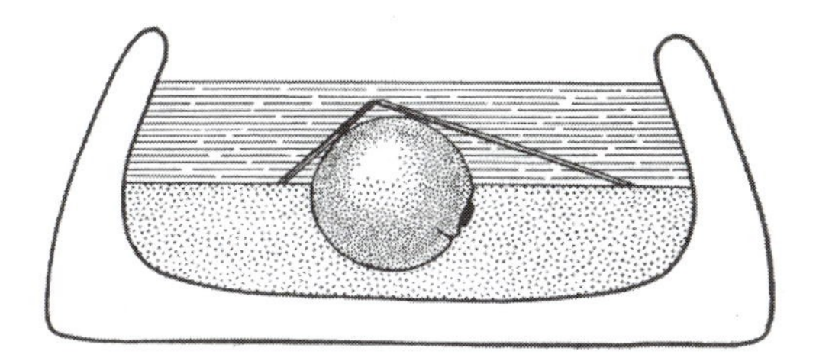

Fig. 6.2 Cross-section through an embryo dyeing chamber, illustrating the depression in which the embryo should be wedged, and the glass bridges that may be used to hold the dye chips in place. (Reprinted from V. Hamburger, *A manual of experimental embryology*, The University of Chicago Press, 1960, with permission of The University of Chicago Press.)

9. The agar should stay in place for 30 minutes.

10. Remove the glass bridge and dyed agar. Gently free the embryos from the modeling clay and transfer them to a clean Petri dish containing fresh RS.

11. Make sketches of your embryos that clearly show where you stained them.

12. Embryos should be observed daily and sketches should be made to illustrate any changes.

13. Embryos will be sacrificed with an overdose of MS-222 during next week's lab session. You will then dissect the animals, which will be about 1 cm in length, with tungsten microneedles (see Appendix 5 (Materials) for instructions) to look for the presence of the dye internally.

QUESTIONS

1. Based on the location of the most intense pigment, draw a map of a gastrula, indicating the size and position of the original dye chip.
2. Did the size of the dye chip have any effect on the survival of the embryo?

7 CELL ADHESION

Cell–cell interactions in sponges and amphibians

1 day

We have been trying to correlate cell movement with tissue formation since the first developmental biologist noticed that individual cells have kinetic properties that might allow morphogenetic events to take place. Roux, in 1894, was the first to actually observe the activities of dissociated cells. Based on the cells' behavior, Roux felt there must be signals secreted by the cells that either attracted or repelled other cells. Holtfreter studied cell-adhesion properties in the late 1930s and 1940s, and came to the conclusion that cell–cell affinities were under temporal (stage-specific) control. This would mean that cells from an early embryo would exhibit general affinity for all other cells from that embryo, whereas those cells from a later stage embryo would only show affinity for cells with a similar tissue fate (Townes and Holtfreter 1955).

The affinity that exists between similar cells is what allows them to form tissues and organs. The measure of this affinity is called adhesion, which is dependent on the intercellular interaction of cell-surface proteins. Generally, when cells have been dissociated, they will begin forming a series of successively stronger bonds with a number of other cells. With time, each cell will find other, more similar, cells with which to form a stronger bond. This process continues until cells are most closely associated with cells of a similar tissue origin (Wilson 1907).

In today's lab period you will be dissociating the cells of embryonic axolotls and mature sponges and observing their ability to reorganize themselves. Sponge integrity is also governed by adhesion molecules (do you remember which molecule is especially important to cell adhesion?), but since they only have three types of cells (epidermal, flagellated, and amoeboid) and no organized tissues, they are less disturbed by dissociation. Sponges can even reproduce by fragmentation of an adult. Vertebrate embryos, on the other hand, with their well-defined tissues and organs, are less able to overcome dissociation, but they are able to reaggregate to some degree and may even go on to develop more or less normally.

TROUBLESHOOTING

- Sponges should be fresh (use within a day or two of acquisition), as they can be difficult to maintain for the inexperienced aquarist.
- Generally, this is a straightforward lab session that rarely gives problems; mainly because we are not trying to manipulate the organisms or keep them alive for very long.

Materials

Depression slides

Artificial seawater (Instant Ocean) $\rho = 1.021–1.025$ g/l

Multiwell culture plates

Cheesecloth (5-inch (2 cm) square/person)

Pipettes

Hemacytometer

Two species of sponge (*Axinella polycapella* (Orange devil's finger) and *Microciona prolifera* (Red beard sponge) are especially good)

Sturdy, blunt forceps

Tongs

Axolotl gastrulae and neurulae

Rearing solution

Calcium-free rearing solution

Rearing solution with 1% KOH (high pH leads to dissociation of adhesion bonds)

0.01% formalin as a fungicide

70% ethanol

Petri dishes

Glass slides

Coverslips

Compound microscope

Dissecting microscope

Axolotl protocol

1. Treat several axolotl embryos of several stages in the formalin solution for 10 minutes.
2. Remove the jelly coat from the axolotl embryos in RS, leaving the vitelline membrane intact.
3. Transfer embryos to RS containing 1% KOH and leave there until their cells have become disarranged (<5 min).
4. Transfer embryo cells to fresh RS (without KOH) and leave for several minutes.
5. Transfer the embryo cells to a new dish of fresh RS.

6. Transfer one or two of the embryos to a third Petri dish and observe after 24 and 48 hours for reaggregation.

7. Transfer the cells of one embryo from step 5 to a slide and cover with a coverslip (you will need to place a short length of hair under two opposite ends of the coverslip or use slivers of coverslips to slightly raise the main coverslip up off the slide).

8. Observe the movement of the cells under 400 × magnification (40 × objective and 10 × eyepieces). You will see Brownian motion, some pseudopodial movement, and aggregation (cells coming together). The movement is not dramatic, and you may need to scan around your slide for a while before you see it. Keep an eye on your cells for 2 hours.

9. You can also try reaggregating your embryos in the calcium-free RS as a comparison. What do you think will happen?

10. You might also try staining the entire exterior of the embryo with a vital dye before dissociating the cells. Look and see if the stained cells reaggregate.

Sponge protocol

1. Obtain chunks of the red and yellow sponges.

2. Wrap each sponge chunk in cheesecloth, submerge in cold seawater, and squeeze with tongs to disperse cells.

3. Transfer cells of each sponge into two multiwell plates containing seawater. For each plate, put cells of the yellow sponge in one well, red in another, and a mixture of the two sponges in a third.

4. Incubate one plate at room temperature and the other in the refrigerator (4 °C). Observe clumping under the dissecting microscope at 24 and 48 hours.

5. Examine another sample of the cells and make an estimate of the ratio of free cells to clumps. This may be easier if you use a hemacytometer.

6. Repeat the estimate at 15, 30, 60, and 120 minutes.

7. Graph your results.

QUESTIONS

1. How much activity do you see in the amphibian cells compared to the sponge cells?
2. How well do amphibian cells reaggregate compared to sponge cells?
3. Do sponges reaggregate faster at 4 °C or at room temperature? Why?
4. How specific are sponge cells about how they reaggregate?
5. Is there any difference between reaggregation in gastrulae and neurulae?

EMBRYOGENESIS

Chick and amphibian development (slide-based)

8

2 weeks

Fertilized chicken eggs are laid at a stage that corresponds to blastula, and within 16 hours (at 38.5 °C) they will begin to gastrulate. Major anatomical structures (head, brain, heart, somites, tail, etc.) are clearly visible within 48 hours of laying, and hatching typically occurs 21 days after laying (see Figs 1.9–1.13 in Chapter 1). Amphibian eggs are fertilized as they are laid, and so development can be observed from the single-cell stage onward. Major anatomical structures are clearly visible within about 70 hours of fertilization (Fig. 8.1).

During today's lab period, you will sketch and familiarize yourself with the development of one organ system (brain, heart, eyes, etc.) or embryonic feature (Hensen's node, body axis flexure etc.) by examining prepared slides containing wholemount embryos (entire embryo has been stained and mounted on a slide) at 18, 24, 36, 48, 72, and 96 hours post fertilization (chick). You will also use cross-sections or sagittal sections of similarly aged embryos (chick and frog) (Figs 8.2–8.9).

Detailed drawings and notes should be made in your lab manual. You should also research the basic developmental patterns of the anatomical structure you studied (primary literature and/or textbooks other than our own). During the next lab period you will present the results of your slide and literature studies to the rest of the class.

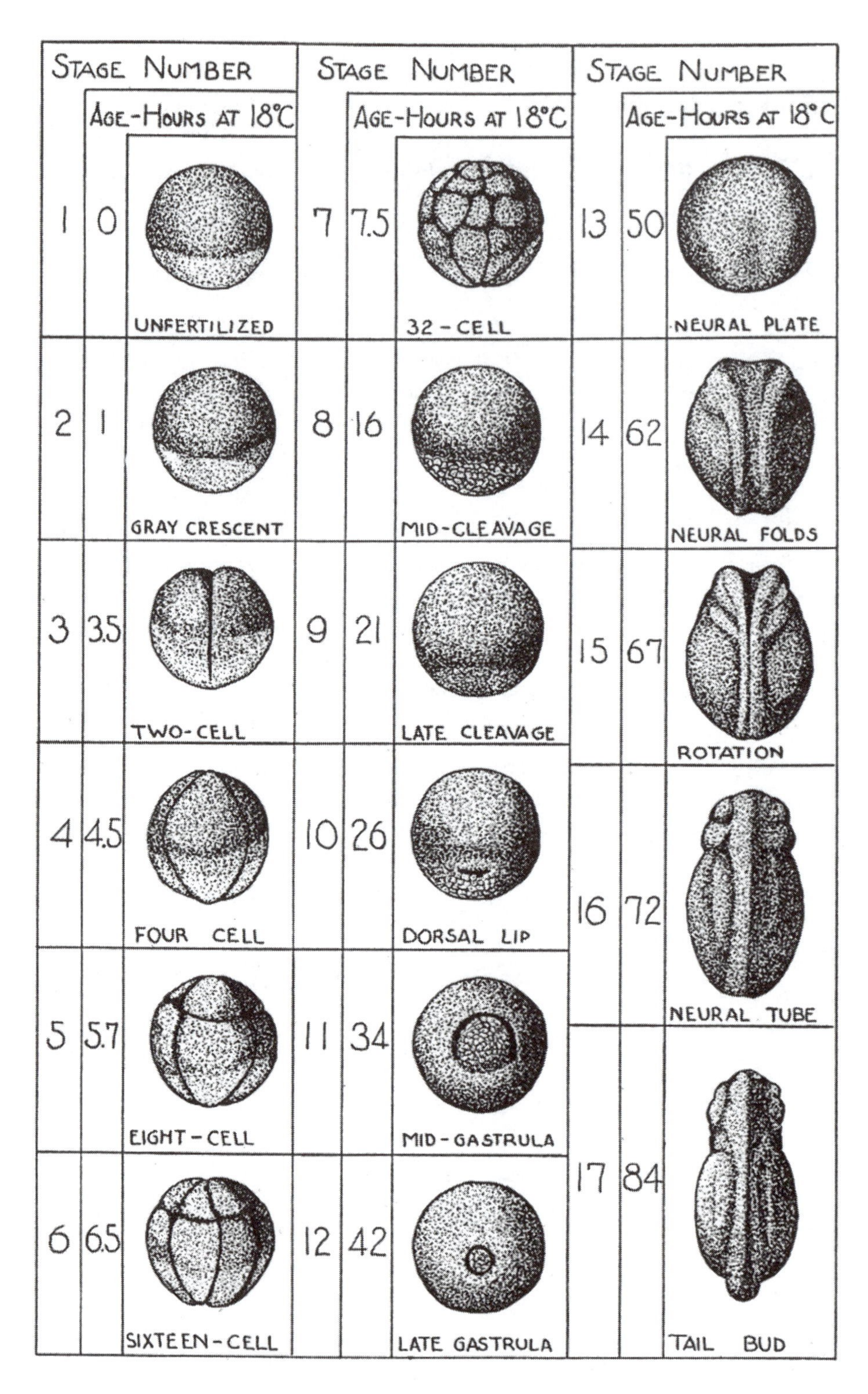

Fig. 8.1 Stages of normal frog (*Rana pipiens*) development. (Reprinted from V. Hamburger, *A manual of experimental embryology*, The University of Chicago Press, 1960, with permission of The University of Chicago Press.)

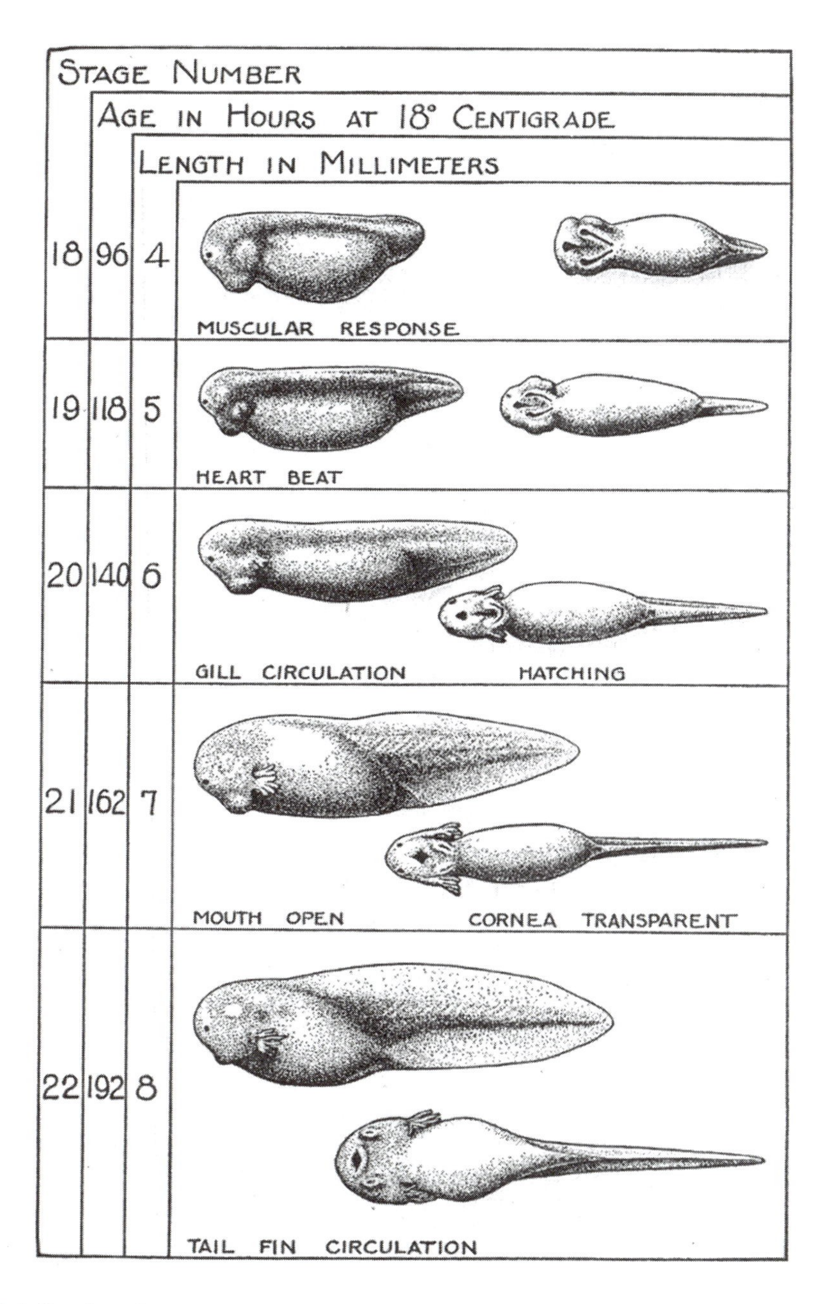

Fig. 8.1 Continued

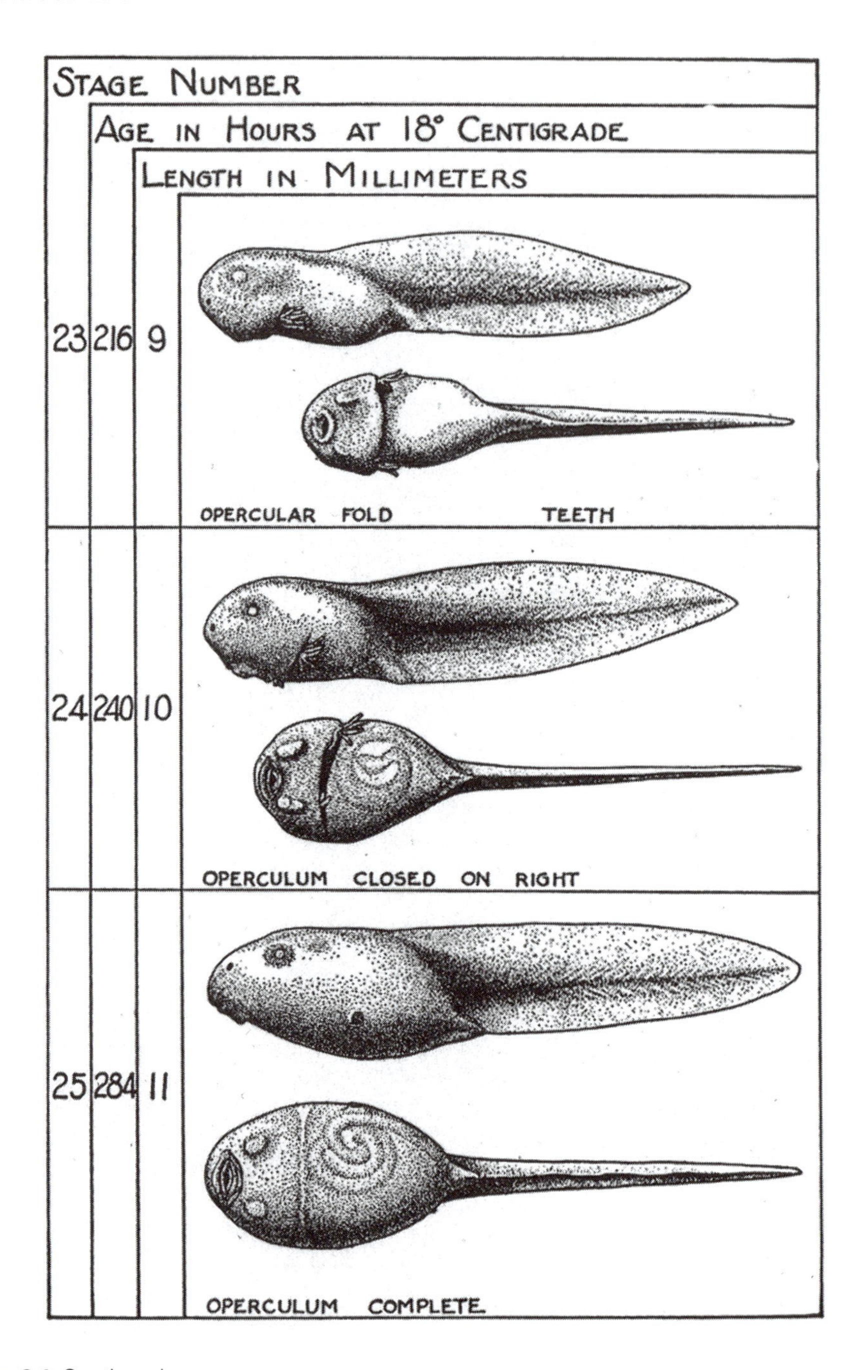

Fig. 8.1 Continued

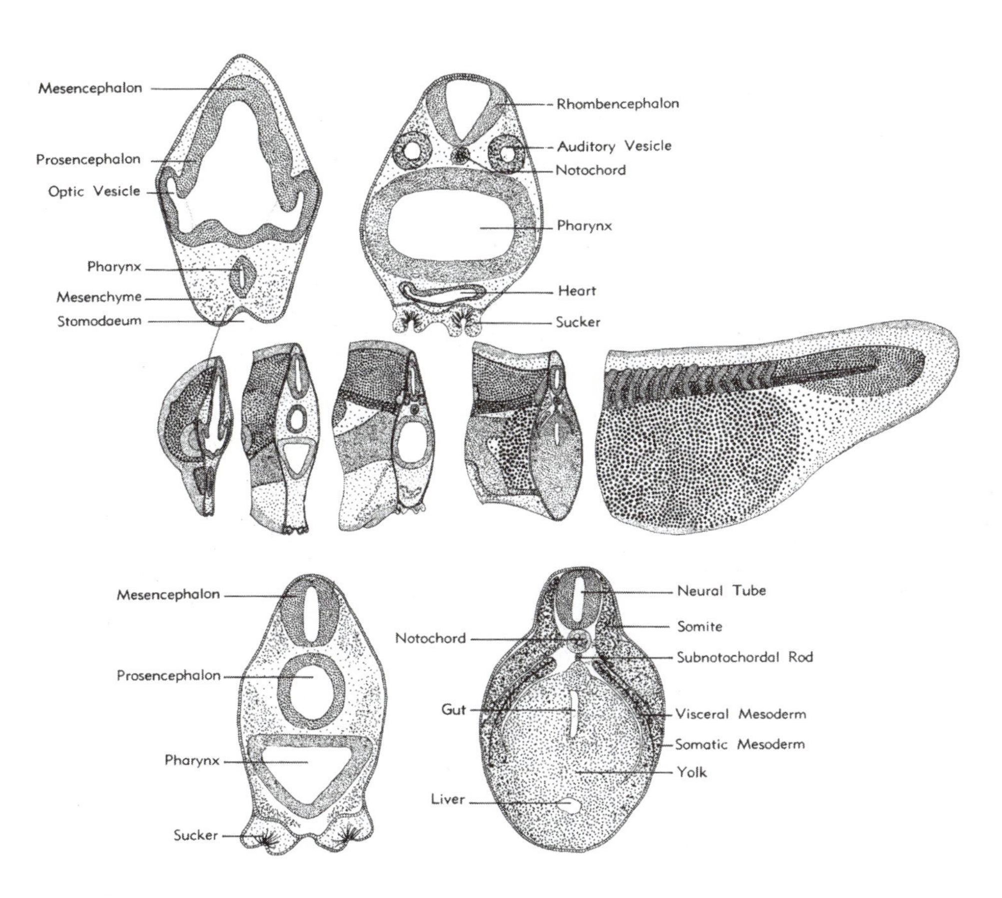

Fig. 8.2 Cross-sections of a 4-mm tadpole larva. (Reprinted from L. E. Downs, *Laboratory embryology of the frog,* Wm C. Brown Co., 1968)

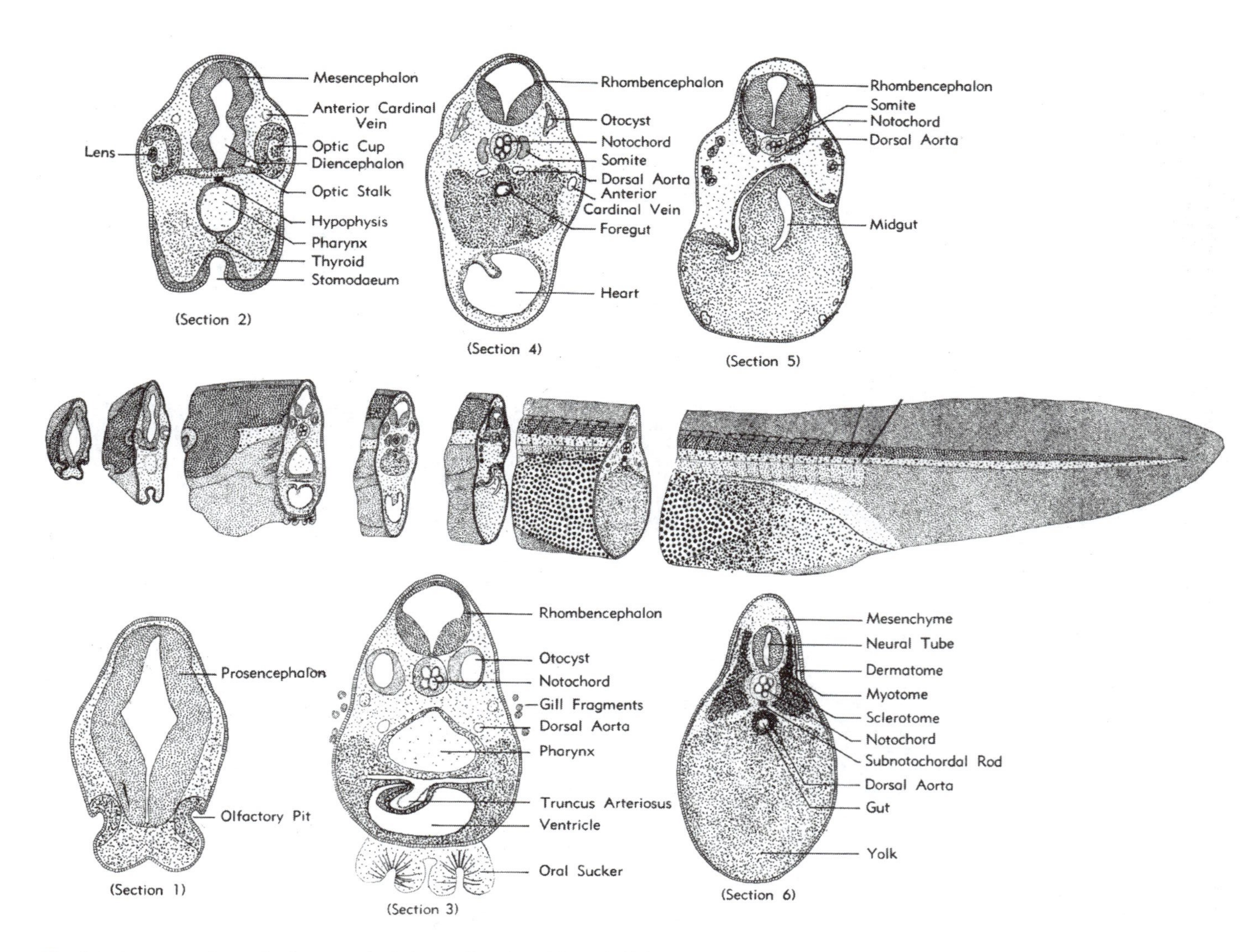

Fig. 8.3 Cross-sections of a 6-mm tadpole larva. (Reprinted from L. E. Downs, *Laboratory embryology of the frog,* Wm C. Brown Co., 1968)

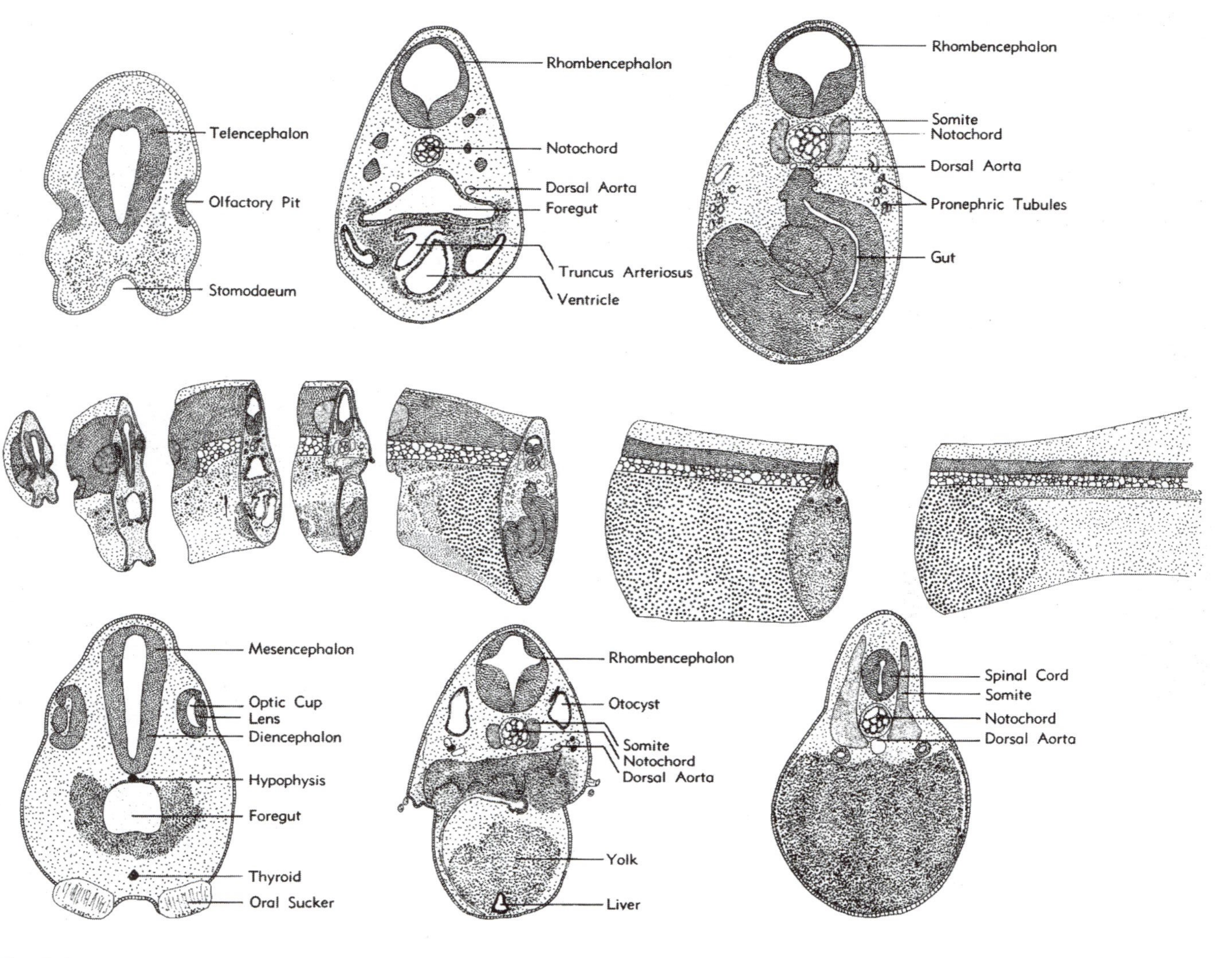

Fig. 8.4 Cross-sections of an 8-mm tadpole larva. (Reprinted from L. E. Downs, *Laboratory embryology of the frog*, Wm C. Brown Co., 1968)

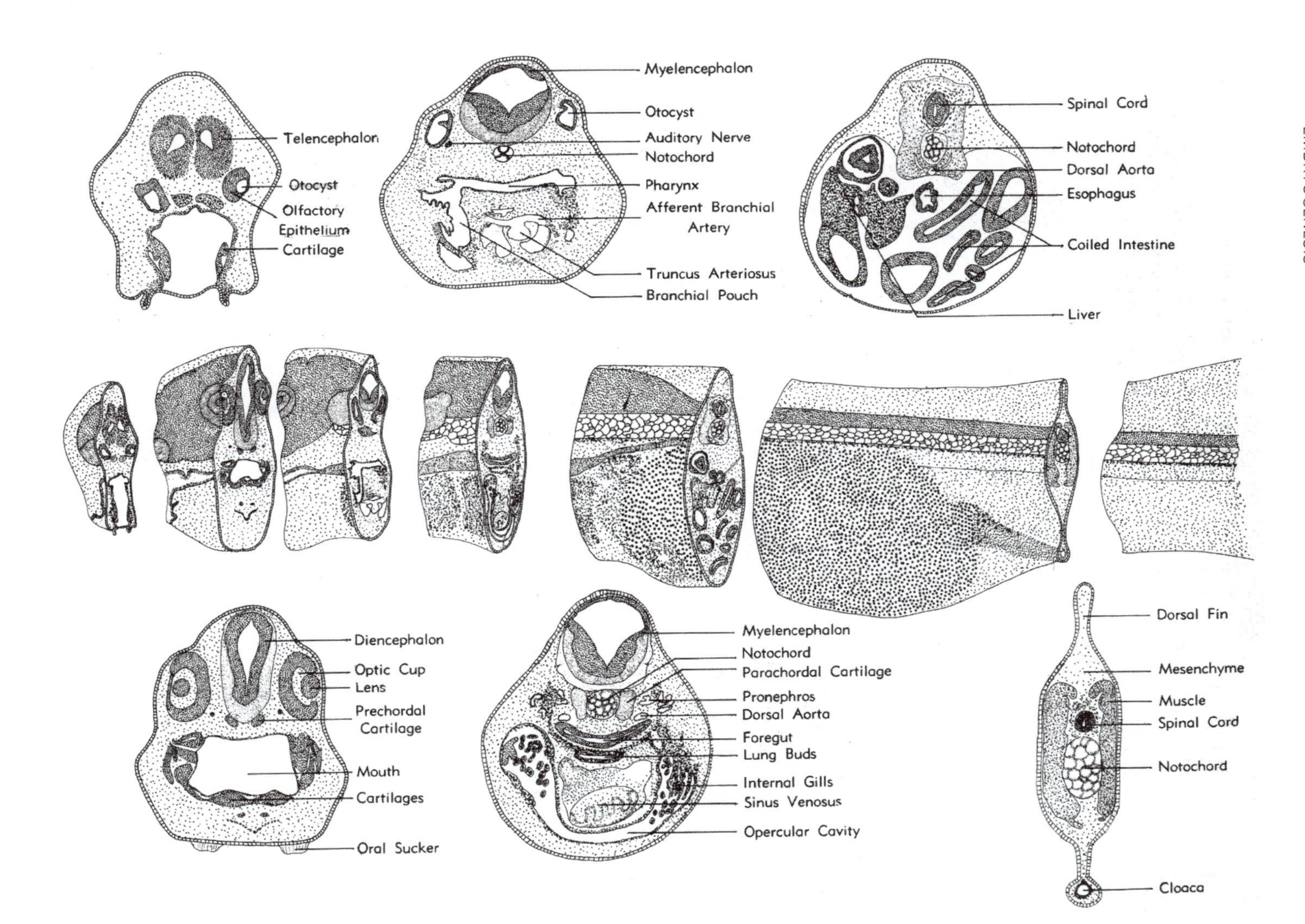

Fig. 8.5 Cross-sections of a 10-mm tadpole larva. (Reprinted from L. E. Downs, *Laboratory embryology of the frog,* Wm C. Brown Co., 1968)

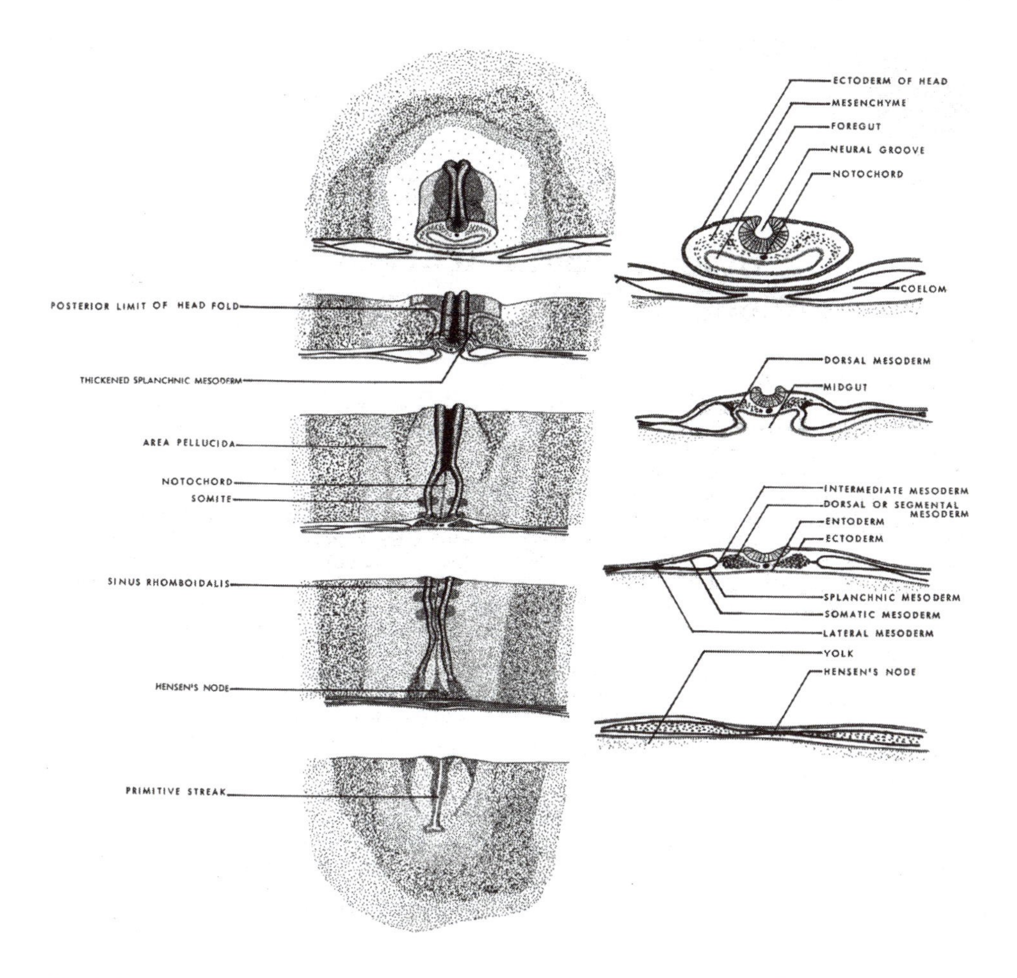

Fig. 8.6 Cross-sections of a stage-8 chick embryo. (Reprinted from L. E. Downs, *Laboratory embryology of the chick*, Wm C. Brown Co., 1963)

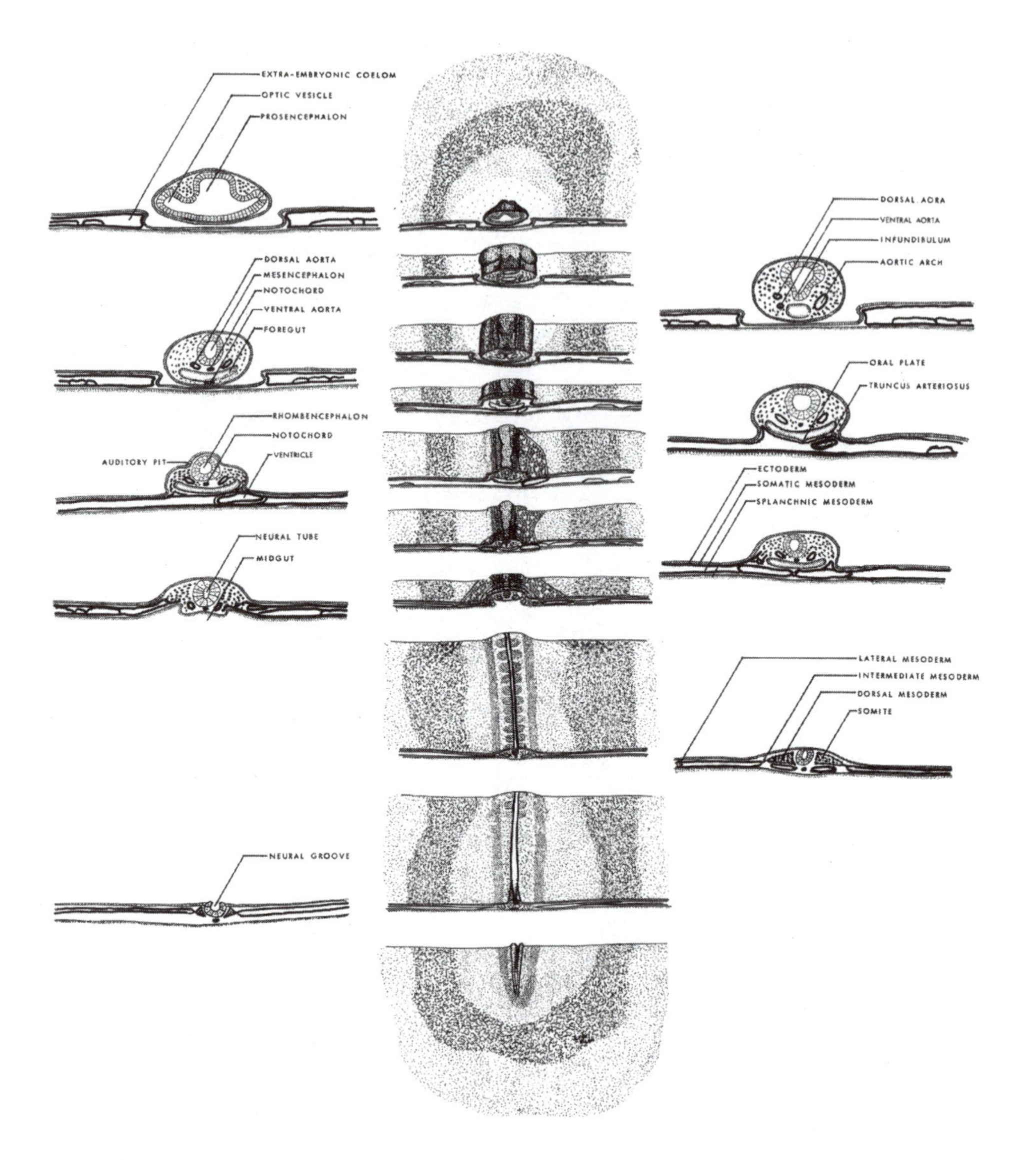

Fig. 8.7 Cross-sections of a stage-10/11 chick embryo. (Reprinted from L. E. Downs, *Laboratory embryology of the chick*, Wm C. Brown Co., 1963)

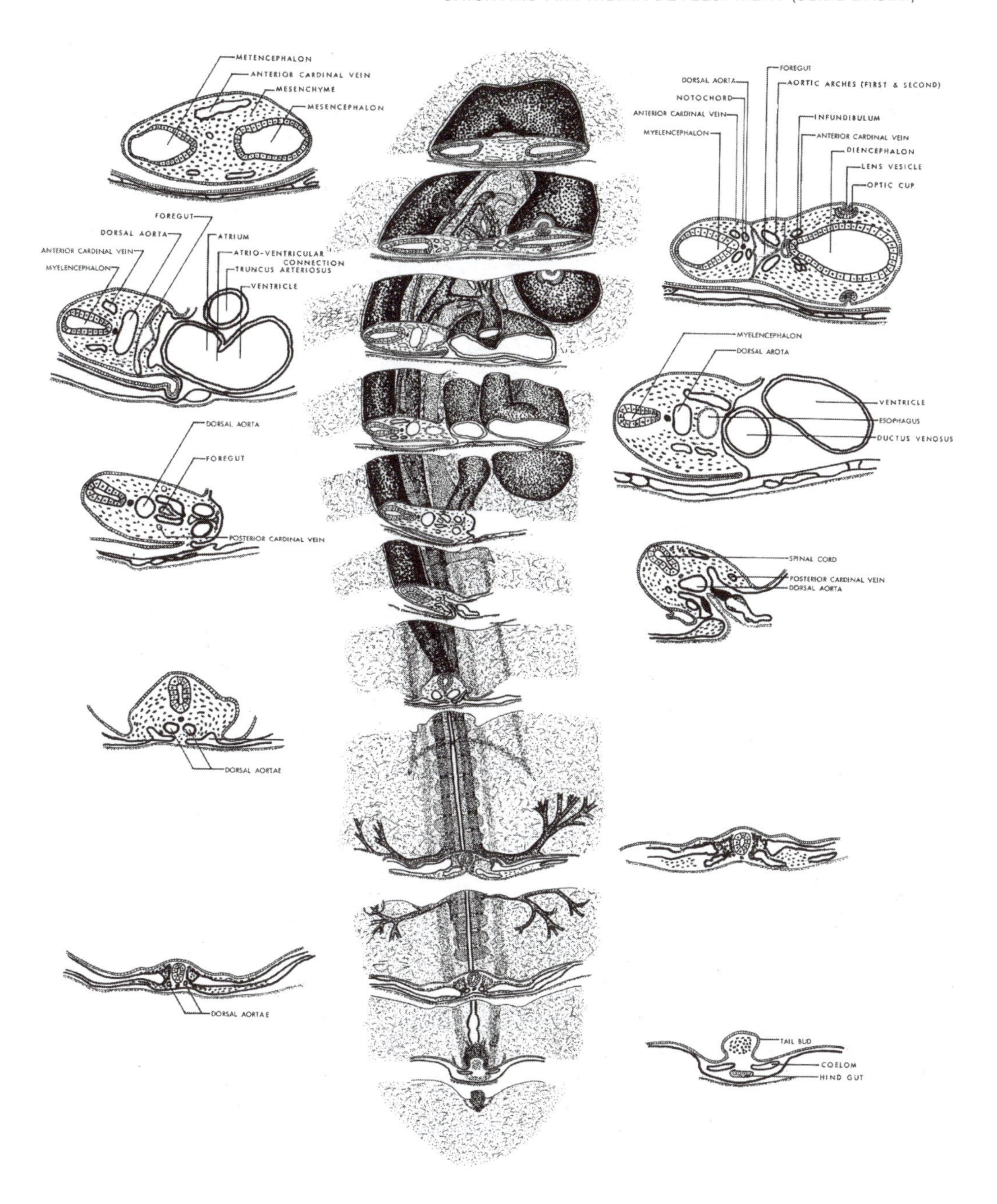

Fig. 8.8 Cross-sections of a stage-12 chick embryo. (Reprinted from L. E. Downs, *Laboratory embryology of the chick*, Wm C. Brown Co., 1963)

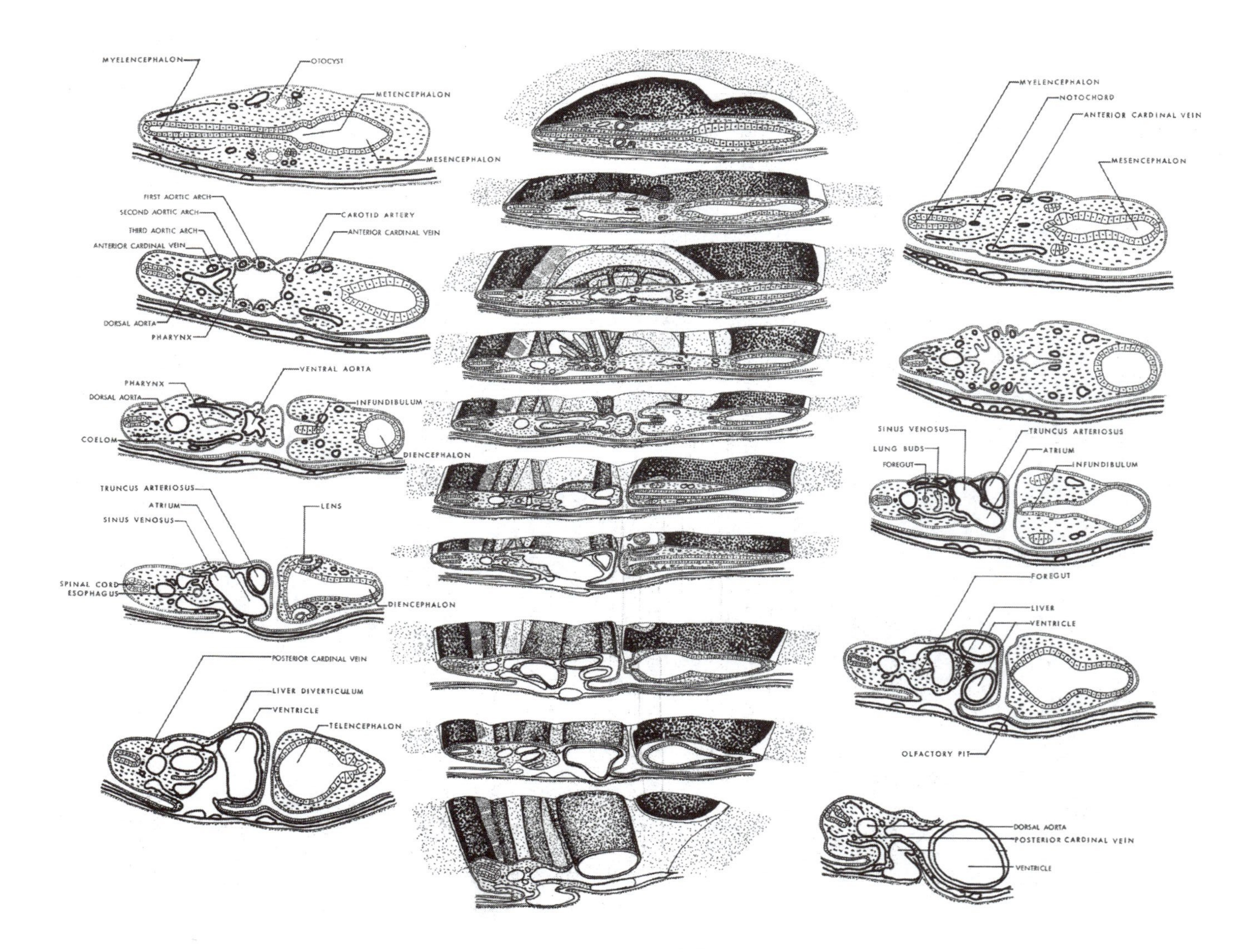

Fig. 8.9 Cross-sections of a stage-18 chick embryo. (Reprinted from L. E. Downs, *Laboratory embryology of the chick*, Wm C. Brown Co., 1963)

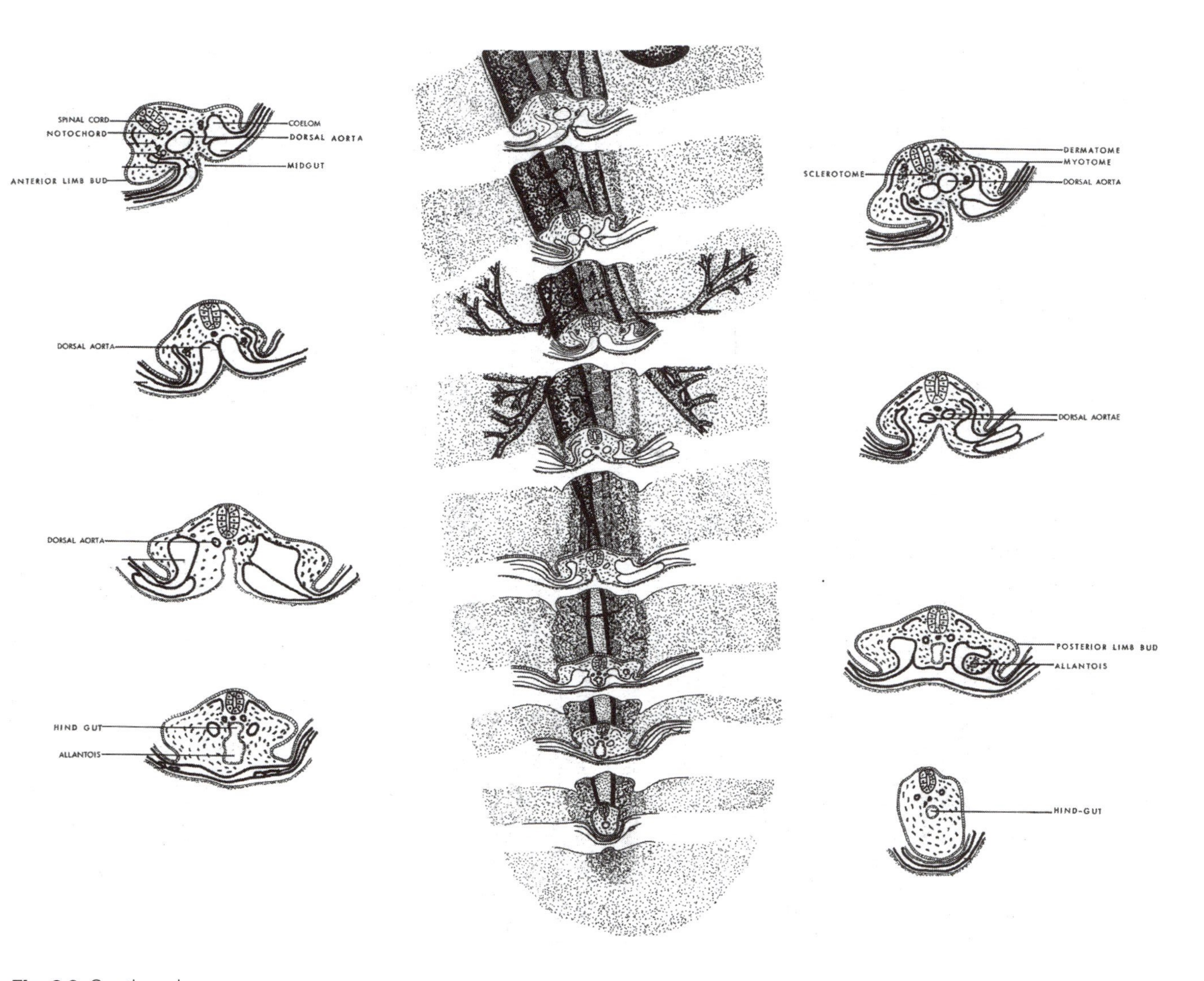

Fig. 8.9 Continued

9 NEURAL INDUCTION

Labeling chick notochord using immunocytochemistry

3 days

Chick gastrulation and neurulation are fascinating because of the anterior–posterior sequence that means gastrulation is ongoing in the posterior region of the embryo, while neurulation has begun in the anterior regions. Neurulation in the chick begins with Hensen's node at the anterior end of the primitive streak, and ends after the node has regressed to the posterior end of the embryo. Hensen's node is similar to the Spemann organizer of amphibians, in that it plays a major role in establishing a visible axis and inducing the nervous system, notochord, somites, and gut endoderm. Cells migrate into the embryo through the primitive streak and Hensen's node; whether cells migrate through the primitive streak or Hensen's node ultimately determines their fate. Cells that migrate through Hensen's node before regression give rise to the head process, pharyngeal endoderm, foregut, and chordamesoderm, which will become the notochord. It is interesting to note that the anterior notochord (near the midbrain) arises from cells that migrated through Hensen's node, but the rest of the notochord, which develops as Hensen's node regresses, arises from cells that migrated through the primitive streak. Hensen's node induces these primitive streak cells to form posterior notochord as it passes by.

The primitive streak is at its longest and Hensen's node contains its greatest inductive potential at about 12 hours postfertilization. By 16 hours, the node has begun to regress towards the posterior end of the streak, and by 28–30 hours it has disappeared. During this short period, radical changes have taken place in the embryo (somites) and the nervous system (neural folds and notochord). Although many more changes will continue to occur, we will focus special attention on the chick notochord. The primary roles of the notochord are the patterning of the ventral neural tube (sonic hedgehog expression) and segmentation of the vertebral column (sclerotome induction). The activity of the notochord is, therefore, restricted to the first 3½–4 days of development, until the entire vertebral column has formed.

Over the next few days, you will use immunohistochemistry techniques to label the active notochord of chick embryos (at varying stages of development) with an antibody against chick notochord (not-1). A series of antibodies and labels can be used to visualize an area of interest immunohistochemically. In today's experiment, you will add a primary antibody (chick not-1 cultured in mice) to your embryo, followed by a secondary antibody directed against the primary antibody's culture source (goat anti-mouse IgG) and conjugated with an enzyme (peroxidase). A chromagen solution, 3,3'-diaminobenzidine (DAB)/peroxide, is then added and reacts with the peroxidase to produce a dark, blue–black stain in the labeled tissue.

WARNING

! *DAB is a carcinogen, so handle with care and gloves. Drop anything that was used with DAB into a bleach solution to inactivate it.*

TROUBLESHOOTING

The degree of notochord staining will depend on the age of your embryo. With older embryos you may need to dissect the spinal cord to find the areas of staining; it will be easier to see in very young embryos that are still in the process of inducing the nervous system.

You may need to leave the embryo in the staining solution for more or less time than is given in the protocol to get the staining at the right intensity.

Materials

Chick saline (3–4 liters per 15 students)

Fertilized eggs (2 or 3 per student), incubated for varying lengths of time (18–72 h)

Fine forceps

Fine-tipped scissors (iridectomy scissors are ideal)

Plastic teaspoons

Large fingerbowls

Small Petri dishes

Glass Pasteur pipettes

4% formaldehyde solution in 0.2 mol/l phosphate buffer (PB)

Phosphate buffer solution (PBS)

Phosphate buffer with Triton-X (PBT) (Triton-X is a detergent that punches holes in the embryo to give the antibodies better access to all parts of the embryo)

Blocking agent (PBT supplemented with normal goat serum (PBT+N)) (goat serum helps prevent, or block, non-specific background staining from overwhelming the embryo and making it difficult, or impossible, to see the notochord labeling)

Primary antibody (not-1)

Secondary antibody (goat anti-mouse IgG peroxidase) (IgG refers to the type of antibody we are using)

1% cobalt chloride ($CoCl_2$) solution

1% nickelous ammonium sulfate ($NiNHSO_4$)

deionized water (dH_2O)

30% hydrogen peroxide (H_2O_2)

Chromagen solution (DAB/peroxide; see Day 3, step 3)

Orbital shaker

Dissecting microscopes

Bleach solution

Notochord labeling protocol (after Schoenwolf 1995)

Day 1

1. Fill a large fingerbowl two-thirds with Chick saline.
2. Crack an egg (18–72 h incubation) on the edge of the fingerbowl and submerge the egg before opening the shell to release the embryo and yolk. Note that these yolks will be considerably more likely to break than the yolk of an egg from the market.
3. Gently maneuver the egg yolk with the plastic spoons or your fingers until the blastoderm is facing up (alternatively, you may carefully pick open the shell overlying the embryo).
4. Use forceps to grip the edge of the blastoderm while you use fine scissors to carefully cut the blastoderm free of the yolk. Use the dissecting microscope if possible.
5. Gently pull the blastoderm away from the yolk.
6. Using a spoon to scoop and the forceps to anchor, transfer the blastoderm to a fresh Petri dish of Chick saline. Remove any yolk still adhering to the blastoderm.
7. Remove the vitelline membrane by gently agitating the blastoderm in the saline. You may need your iridectomy scissors to trim membranes that don't separate easily.
8. Transfer the blastoderm into a large Petri dish containing fresh saline. Make sure the blastoderm is face up and spread out.
9. Use a Pasteur pipette to slowly siphon off the saline, so that the blastoderm settles smoothly onto the bottom of the dish.
10. Add 4% formaldehyde solution (in 0.2 mol/l PB) drop by drop, directly onto the embryo, until the embryo is completely covered.
11. Allow the specimen to fix for 1½–2 hours.
12. Wash the specimen for 5 minutes in PBS. Repeat.
13. Wash the specimen for 5 minutes in PBT. Repeat.

14. Wash the specimen for 30 minutes in PBT on the orbital shaker.
15. Wash the specimen for 30 minutes in PBT+N in the refrigerator.
16. Remove the PBT+N and add the not-1 (primary antibody) solution (diluted 1:10 in PBT+N).
17. Incubate overnight in the refrigerator (or in the refrigerator for a week).

Day 2

1. Remove the not-1 solution by washing the specimen twice in PBT (5 minutes each wash).
2. Wash the specimen for 30 minutes in PBT on the orbital shaker. Repeat three times.
3. Wash the specimen for 30 minutes at room temperature in PBT+N on the orbital shaker.
4. Remove the PBT+N with a Pasteur pipette and add goat anti-mouse IgG peroxidase diluted 1:200 in PBT+N (secondary antibody solution).
5. Soak the specimen in the secondary antibody solution overnight in the refrigerator (or in the refrigerator for a week).

Day 3

1. Remove the goat anti-mouse IgG peroxidase solution by washing the specimen for 5 minutes in PBT. Repeat two times.
2. Wash the specimen for 30 minutes in PBT on the orbital shaker. Repeat three times.
3. Make DAB solutions while the specimen is on the orbital shaker:

   (a) DAB/PBT solution: 2 ml PBT + 1 ml DAB/PBT stock + 75 μl $CoCl_2$ + 60 μl $NiNHSO_4$ (nickelous ammonium sulfate);

   (b) Peroxide solution: 2.5 ml dH_2O + 25 μl 30% H_2O_2.
4. Remove the PBT, add the DAB/PBT solution, and allow to react for 10 minutes.
5. Prepare the staining solution by adding 1 ml of the DAB/PBT solution to 12 μl of the peroxide solution.
6. Remove the DAB/PBT solution and replace with staining solution. Watch the embryo under a dissecting microscope until the desired level of staining is observed.
7. Stop the reaction by removing the staining solution and replacing it with PBS. Store in the refrigerator.

QUESTIONS

1. During which developmental stage is not-1 most highly expressed?
2. Do different stages express not-1 in different areas of the body? Why?
3. At which stage is not-1 no longer expressed?

GAMETOGENESIS

Comparison of animal and plant gametogenesis

10

1–2 days

It is important to remember that gametogenesis actually begins in the developing embryo, arrests (or slows) as spermatogonia or oogonia before birth, and then resumes at the onset of puberty. Spermatogonia never completely stop dividing prior to puberty in frogs; however, at puberty, they will start entering meiosis. These primary spermatocytes will each form two secondary spermatocytes. Secondary spermatocytes go through another meiotic division to form spermatids (haploid). Spermatids mature into sperm as they move through the testes. Sperm maturation is pretty much the same in all vertebrates (Figs 10.1–10.3).

Oogonia become oocytes when they enter meiosis. Instead of moving through all meiotic stages, they arrest in diplotene of prophase I in order to go into a period of growth and differentiation (lampbrush chromosomes form, indicating high levels of RNA synthesis). This activity is necessary for the presumptive egg to acquire the volume of protein needed for early embryonic development. Also during this stage, yolk is extracted from the mother's bloodstream (yolk precursors are made in the liver) in a process called vitillogenesis. After vitillogenesis, the oocyte waits for hormonal signals (luteinizing hormone from the pituitary gland stimulates ovarian follicle cells to produce progesterone) to trigger oocyte maturation and a return to meiosis. Once meiosis is complete, the oocyte becomes an egg.

Chick oogonia stop mitotic division just before hatching. At this point the cells become larger and initiate meiosis (primary oocytes), after which they are surrounded by a layer of follicle cells (Figs 10.4 and 10.5). All this occurs in the left ovary only (chickens, for whatever reason only they know, do not use the right ovary to produce eggs); follicles will remain dormant until puberty, when follicle-stimulating hormone induces yolk synthesis in the liver. After vitillogenesis, luteinizing hormone stimulates ovulation of the largest secondary oocyte into the oviduct, where it is packaged (albumin and shell) for laying.

Plant development (especially pollen) was described in Chapter 5. Now you will have the opportunity to view the internal structure of the *Lilium* carpel. The carpel contains the ovule, which may have originally produced as many as four megaspores but usually contains a single megaspore by the time the flower approaches maturity (Fig. 10.6). This megaspore develops into a megagametophyte, which then goes through three divisions to produce a variety of cell types, including an egg. The entire ovule is surrounded by integumentary layers, except for a small area called the micropyle, through which the pollen tube will enter. When the pollen tube reaches the ovule, one sperm cell will fuse with the egg to form a zygote and the other will fuse with polar nuclei to form the triploid endosperm. Endosperm nourishes the developing

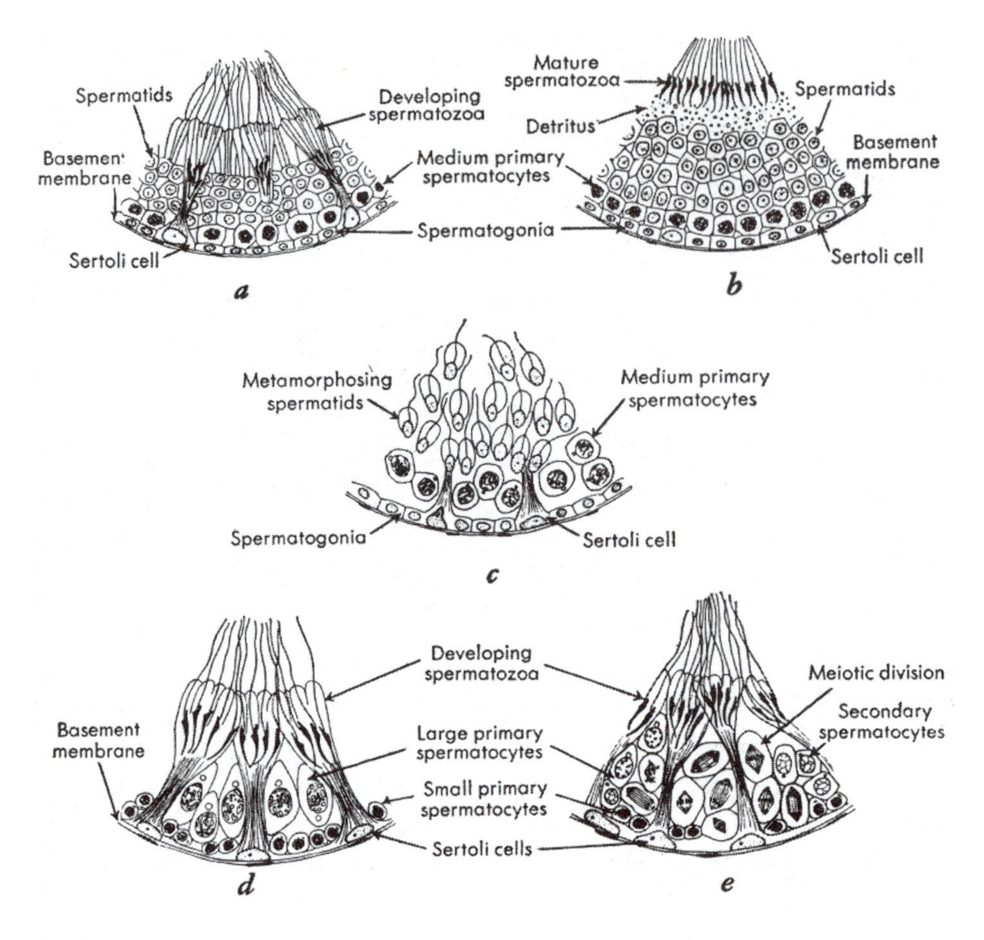

Fig. 10.1 Spermatogenesis in the rat. Figures a–e illustrate the many spermatogenic waves of development occurring in different parts of the same seminiferous tubules ('a' represents the earliest stages of development, and 'e' the latest). (Reprinted from R. M. Eakin, *Vertebrate embryology, University of California Syllabus Series No. 376*, 1958, with permission of the University of California Press.)

embryo (e.g. meat and milk of coconuts or the kernel of corn) (Fig. 10.7). Angiosperms, therefore, essentially go through a double fertilization!

Materials

Compound microscope

Slides of testes and ovaries of three animals and the reproductive organs of a plant

Colored pencils

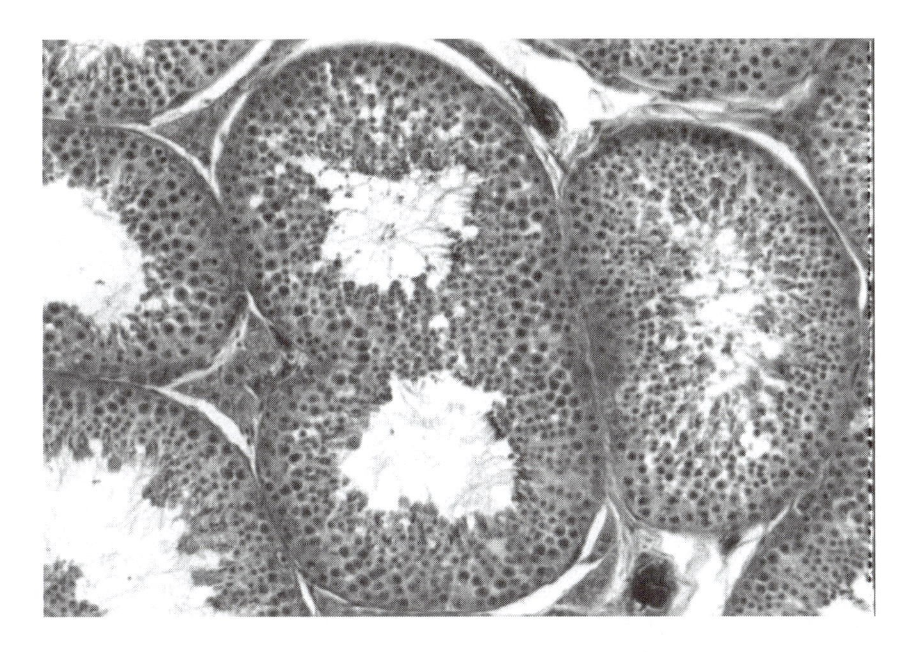

Fig. 10.2 Cross-section through a cat testis, showing the seminiferous tubules. (100×)

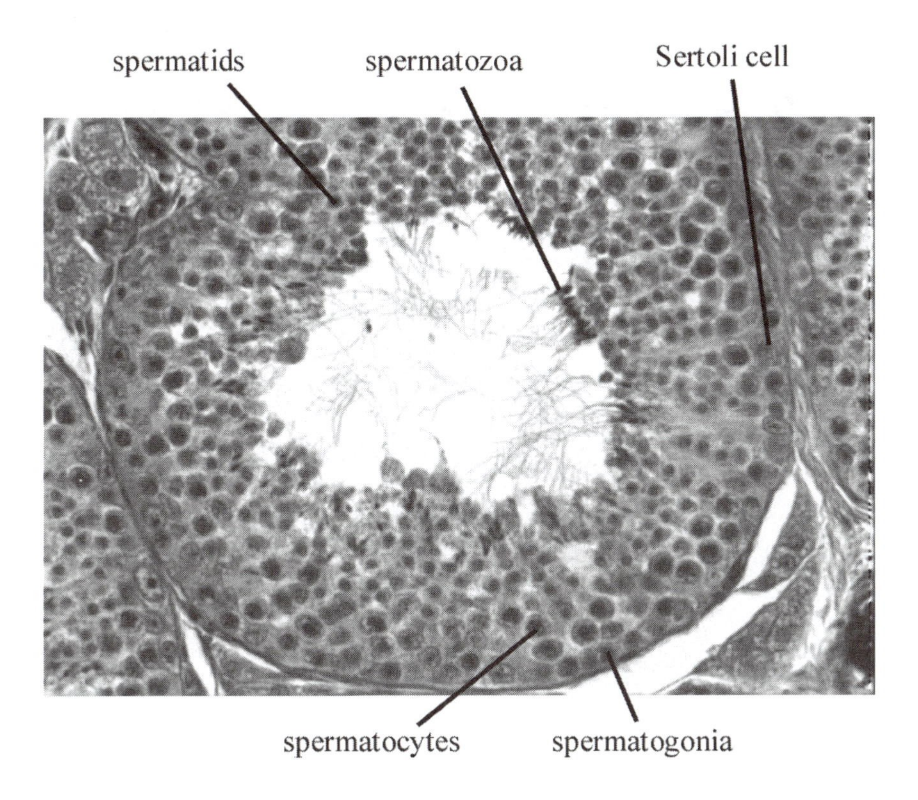

Fig. 10.3 Close-up of a cross-section through a cat testis, illustrating spermatogenesis. (400×)

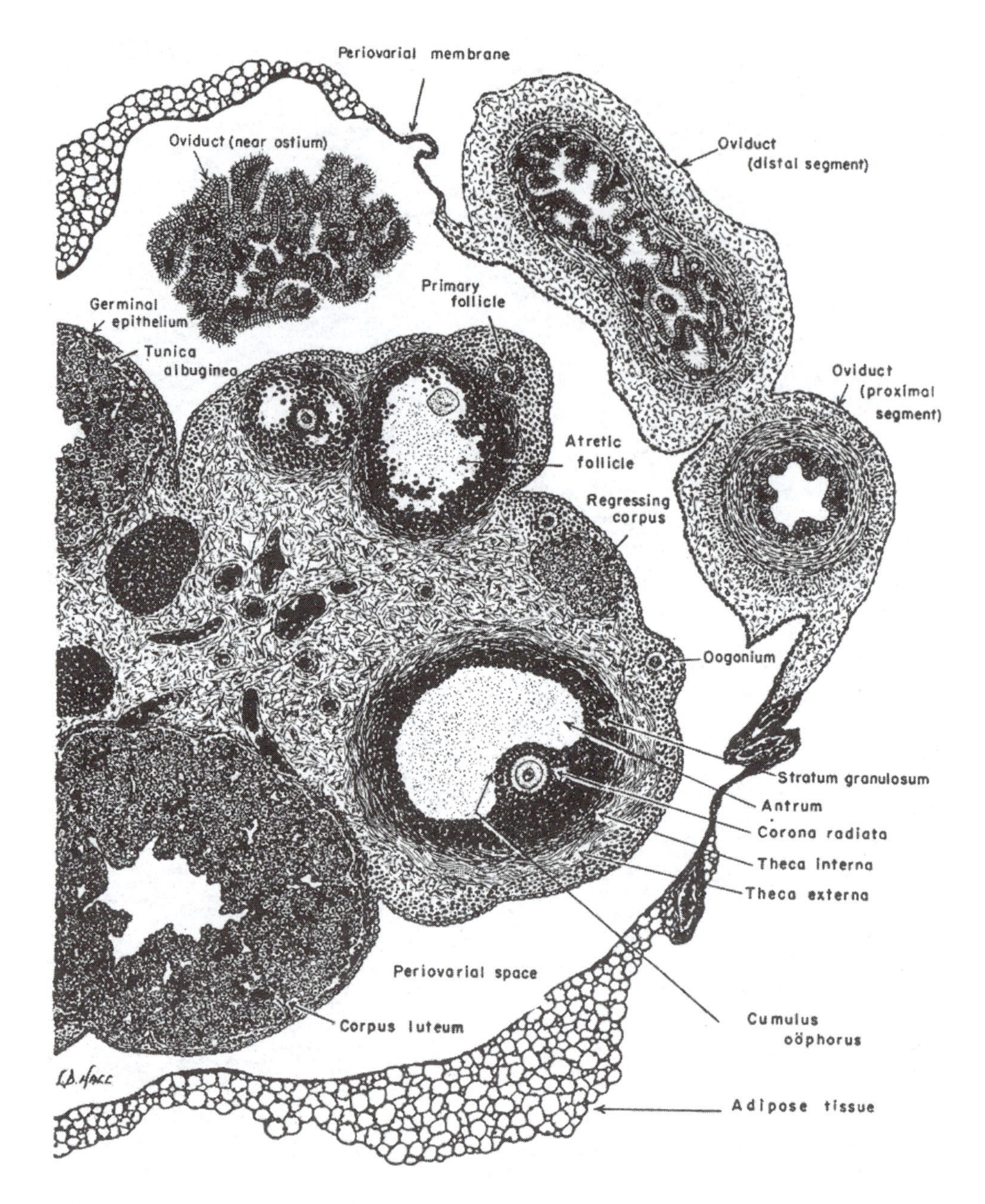

Fig. 10.4 Cross-section through a rat ovary, illustrating the stages of oogenesis. (Reprinted from R. M. Eakin, *Vertebrate embryology, University of California Syllabus Series No. 376*, 1958, with permission of the University of California Press.) (~40×)

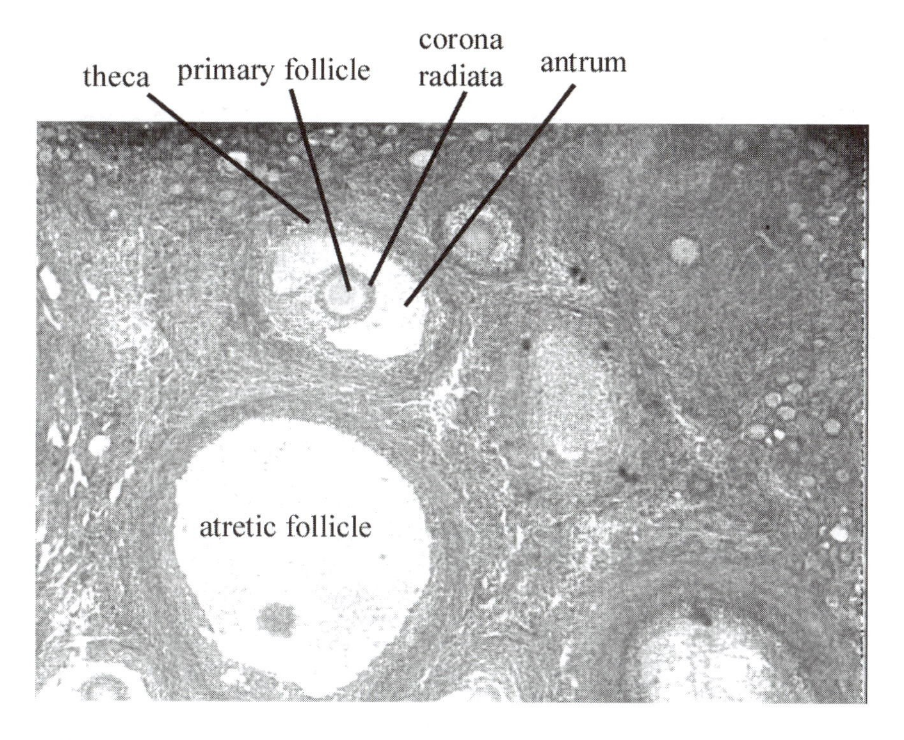

Fig. 10.5 Cross-section through a cat ovary, illustrating oogenesis. (40×)

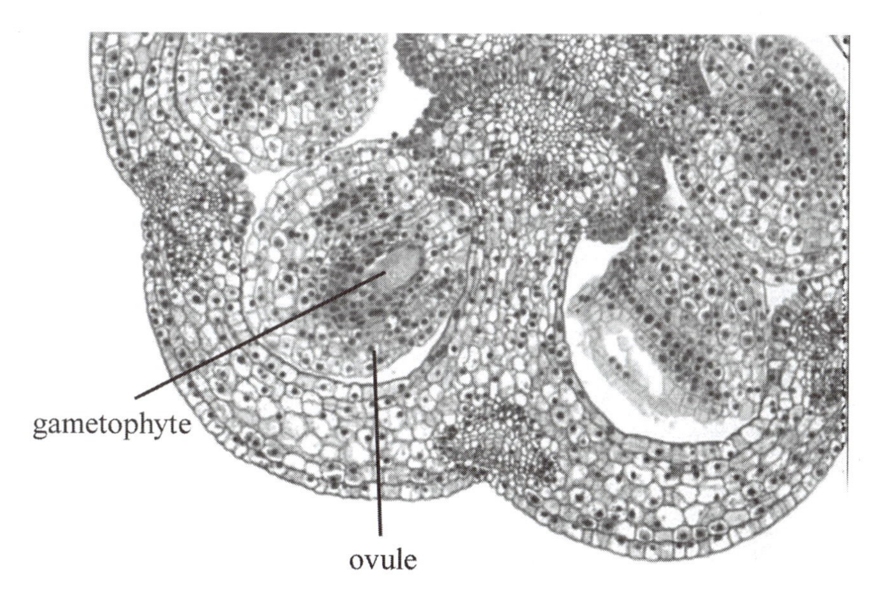

Fig. 10.6 *Lilium* ovary, illustrating the position of the ovule and gametophyte. (40×)

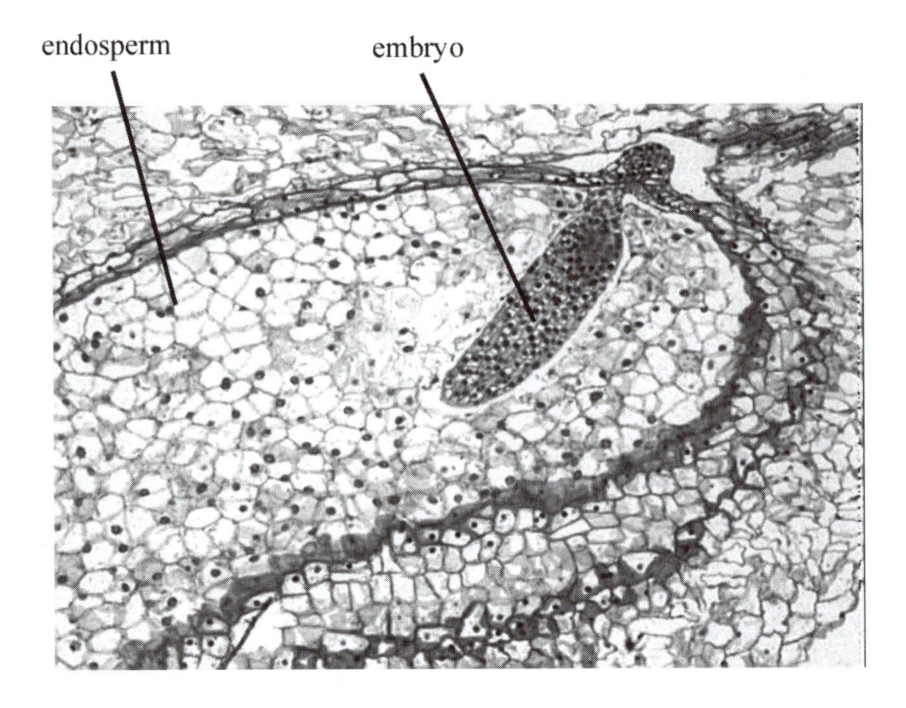

Fig. 10.7 Close-up of a *Lilium* ovule, containing a young embryo and its nutrient endosperm. (40×)

Protocol

1. Sketch and compare the testes of three animals (e.g. out of human, cat, frog, chick).
2. Label seminiferous tubules, spermatogonia, spermatocytes, spermatids, and spermatozoa in each drawing.
3. Sketch and compare the ovaries of three animals you have chosen (e.g. out of human, cat, frog, chick).
4. Label the primary follicle, mature follicle, oocytes, and corpus luteum in each drawing.
5. Sketch the pollen and ovary of three flowering plant along with their basic flower structure.
6. Label the stigma, style, ovary, anther, filament, stamen, micropyle, nurse cells, and endosperm.

QUESTIONS

1. Compare and contrast sperm and ovum morphology and development among the three animal species you looked at.
2. Try and find similarities between animal and plant gametogenesis.

11 REGENERATION

Morphallaxis and epimorphosis

2 weeks

People have known since the 1700s that the flatworm planaria (common in ponds and streams) will regenerate lost pieces. However, it wasn't until the late 1800s that Morgan (1898) conducted a thorough study of their regenerative abilities. Quite a few organisms (*Hydra*, oligochaete worms, lizards) are able to replace body parts that have been lost traumatically by utilizing signal gradients and turning back the molecular clock in cells next to the wound. All forms of regeneration are strongly governed by signal gradients that have established some sort of polarity (anterior–posterior, proximal–distal, etc.), and that dictate the formation of appropriate body parts. In other words, for example: wherever signal X is highest, a head will exist; and wherever this signal is lowest, will be the tail. This way, if you cut off the head of something like *Hydra* or *Lumbriculus*, the area of the neck will now have the highest signal X concentration. The gradient directs cells to respecify themselves into a new conformation and function. This type of regeneration, in which one part of the body shifts its identity to that of a missing piece, is called morphallaxis. A distinctive feature of morphallaxis is that there is a real change in the size of the organism from injury to complete regeneration. If you cut the oligochaete worm *Lumbriculus* in half, it will regenerate the missing pieces, but it will still be the size of half a worm.

In contrast to morphallaxis, the second category of regeneration, epimorphosis, does involve new growth. Mature cells near the wound site will dedifferentiate to the point when they can start to proliferate, forming a blastema (much like the plant callus). These relatively undifferentiated blastema cells will give rise to a variety of cell types on the basis of positional gradients. If you cut off the head of a planarian worm, it will grow a new one; and since it **is** new growth, the resulting animal will be as long as a normal planaria (Fig. 11.1). Blastemas create appropriate structures due to *Hox* gene control. If you treat an amphibian blastema with retinoic acid, you will alter the structures it produces (e.g. if the blastema would have made wrist and fingers, then after retinoic acid treatment it will make a radius and ulna in addition to fingers and wrist).

In today's lab session you will initiate morphallaxis in the worm *Lumbriculus* by cutting its body in half. Epimorphosis will be initiated in one or more species of planarian worms (different species have different regeneration rates and abilities). During the following 2 weeks, you will watch for the formation of a blastema in the planaria and the reappearance of missing organs in *Lumbriculus* as it respecifies its tissues.

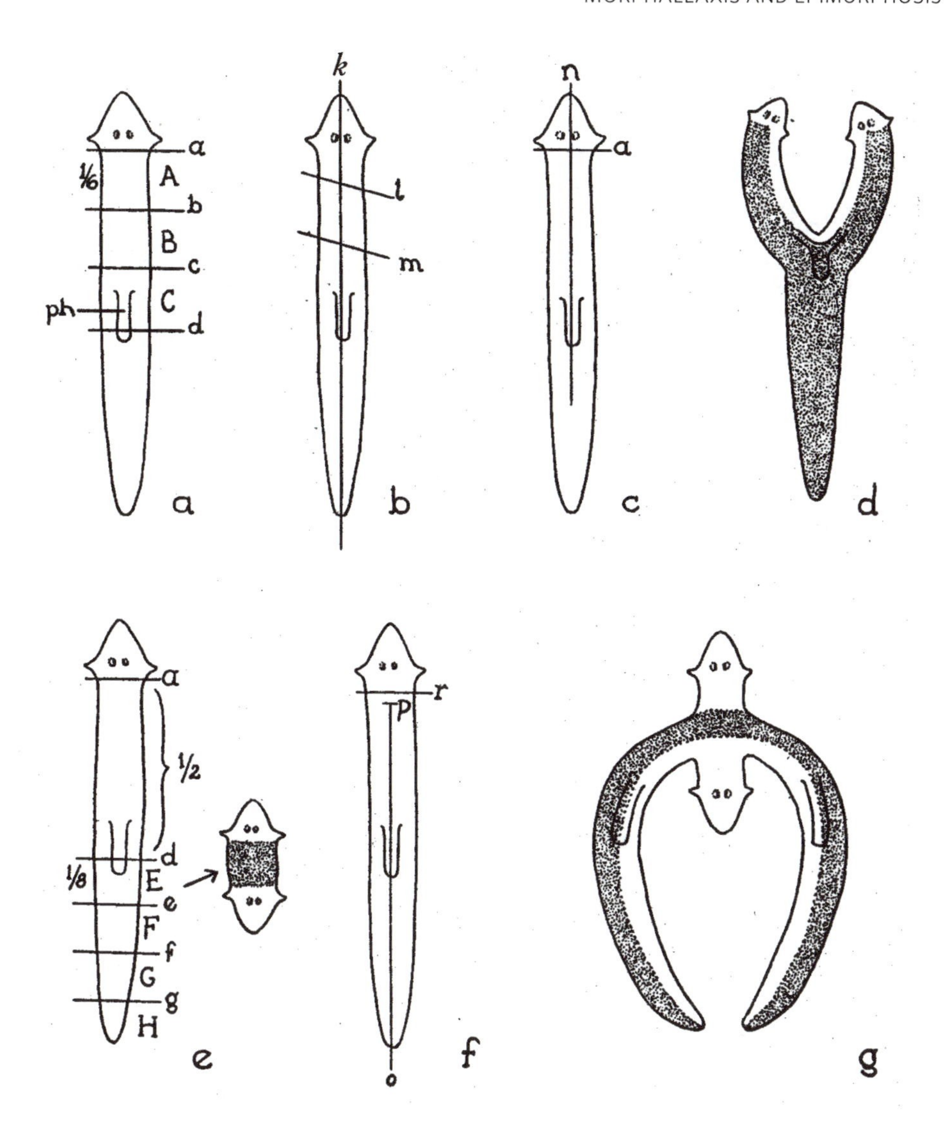

Fig. 11.1 Examples of epimorphic regeneration in planaria, illustrating the types of cuts that can be made and the regenerative patterns that will result. A, B, C, E, F, G, H indicate sections of planaria. Lower case letters show where cuts were made; m and l are oblique cuts, k indicates that the section was made in the medial plane. The fractions indicate how much of the total length of the planaria was cut. Fig. (d) is the result of cuts n and a in Fig. (c), Section E in Fig. (e) gives rise to the 2-headed planaria piece, and Fig. (g) is the result of cuts r, p, and o; dotted areas are old tissue, white is regenerated. The abbrev. ph = pharynx. (Reprinted from V. Hamburger, *A manual of experimental embryology*, The University of Chicago Press, 1960, with permission of The University of Chicago Press.)

TROUBLESHOOTING

- Contamination is one of the major pitfalls of this experiment as it will lead to infection and death of your animals. If your animals disintegrate, you need to review your sterile embryo techniques.
- Don't feed the planaria or *Lumbriculus* during the week prior to experimentation.

Materials

Lumbriculus

Planaria (*Dugesia dorotocephalus* and/or *D. tigrina*)

Artificial spring water

70% ethanol

Glass slides

Large Petri dishes

Ice

Scalpel (with new blade)

Pipettes

Small paintbrushes

Dissecting microscope

Planaria protocol

1. Sterilize all equipment in 70% ethanol.
2. Fill a Petri dish with ice and settle a slide onto it. Place a planaria on the slide (the cold will slow it down and anesthetize it) and wait for it to relax (spread out).
3. Determine in which part of the planaria you wish to observe regeneration, and quickly make a cut.
4. Transfer the pieces to a Petri dish of spring water using the paintbrush.
5. If you are trying to grow a two-headed or two-tailed planaria, you will need to maintain the cut over the next few days, as it will tend to grow back.
6. Change half the water every other day and remove any fragments or dead animals. Keep your planaria culture in a cool, dark location.
7. Sketch the animals every 2 days for 2 weeks; be on the lookout for development of the blastema and regrowth of the missing pieces.

Lumbriculus protocol

1. Follow the planaria protocol through step 4.
2. Change half the water every other day and remove any fragments or dead animals.
3. Sketch the animals every 2 days for 2 weeks. Keep an eye on the internal anatomy of the worms and see how it changes with time.

QUESTIONS

1. Which species of planaria regenerated fastest?
2. Did one species appear to regenerate better than the other?
3. Did *Lumbriculus* and planaria regenerate at the same rate?

12 METAMORPHOSIS

Drosophila imaginal discs

2 days

One of the many reorganizational processes that occur during the metamorphosis of *Drosophila* is the evagination of imaginal discs into adult structures such as eyes, antennae, wings, legs, halteres, and genitals. Imaginal discs are self-contained clusters of epithelial cells in insect larvae that form early in embryonic development from small invaginations of ectoderm. As the embryo, and then the larva, grows, so will the imaginal discs. Although there are definite changes in imaginal disc size as the larva grows, the change in shape and conformation of the original pouch is most intriguing. Because growth occurs within a confined space, a fully developed larval imaginal disc resembles a flattened coil or collapsible camping cup (Figs 12.1–12.3).

There are a total of nine pairs of imaginal discs (labrum, labial, antennal, eye, wing, three legs, and halteres) plus a single gonadal disc (see Fig. 1.3 in Chapter 1). The identity of these discs is established by the anterior–posterior pattern of homeobox genes (HOM complex). Imaginal discs are quite small during the 1st- and 2nd-instar stages, but begin to proliferate during the 3rd-instar stage. During pupation, the hormone ecdysone is secreted and initiates evagination of the imaginal discs. The centralmost portions of the discs will evaginate first and form the most distal part of the resulting structure. Once the discs have been exposed to ecdysone, they will telescope out into adult structures within 24 hours.

Today you will dissect imaginal discs out of 3rd-instar *Drosophila virilis* larvae, expose discs to different concentrations of ecdysone, and monitor the results over 24 hours. You may also artificially evaginate some imaginal discs by exposing them to a proteolytic enzyme solution. Although you may have some difficulties in finding all imaginal discs, it is quite easy to locate the most anterior discs (especially the legs and eye-antenna).

TROUBLESHOOTING

Can't find discs

These are most obvious near the brain, so if you can't actually distinguish the discs, transfer the brain and surrounding tissues into the treatment solutions. Odds are that you will have transferred some discs as well.

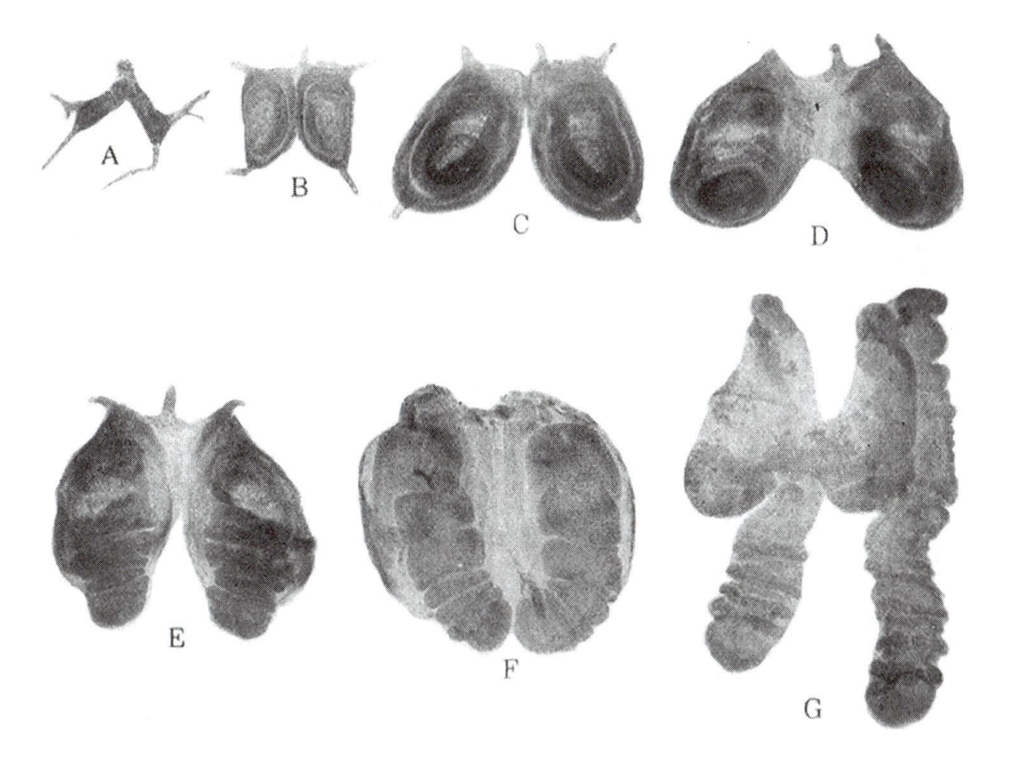

Fig. 12.1 Development of leg imaginal discs. (Taken from M. Demerec, *The biology of* Drosophila, 1950)

Didn't transfer any of the discs you found

Practice with bits of debris before you begin the experiment to get a feel for the microneedles. All parts of the transfer should be done under the dissecting microscope so you can see what you are really doing.

Nothing telescoped

- You might not have had any discs in your treatment chamber.
- The enzyme might have become inactivated.
- Never make the solution up far in advance.

Materials

Tools

Drosophila virilis 3rd-instar larvae (these will be crawling up the incubation chambers)

Small paintbrush (for removing laïrvae from the incubation chamber)

Five, sterile 34-mm Petri dishes or a multiwell tissue culture plate (you will need four of these for the ecdysone protocol and one for the trypsin protocol)

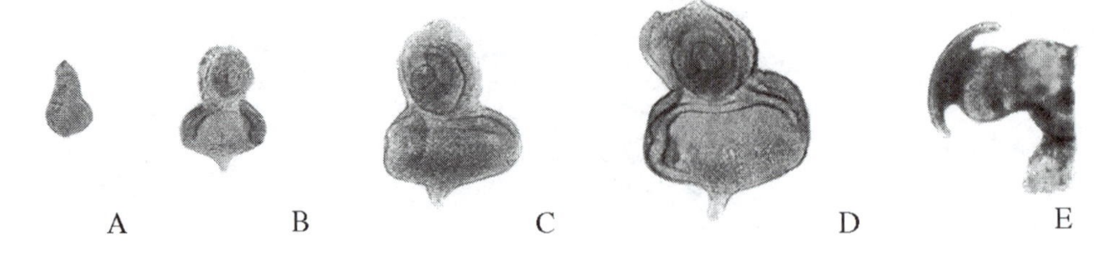

Fig. 12.2 Development of eye-antenna imaginal discs. (Taken from M. Demerec, *The biology of* Drosophila, 1950)

One, sterile Petri dish for dissecting larvae

2 pairs of sterile forceps

Micropipettors (1–5 μl and 200–1000 μl)

Tungsten microneedles

Sterile 2–5 ml glass pipettes

Compound microscope

Dissecting microscope

Ordinary and depression microscope slides

Solutions

Sterile Insect saline

70% ethanol

95% ethanol

Schneider's Insect Medium supplemented with 10% fetal calf serum (containing growth factors)

Ecdysone solution (1 mg ecdysone/1 ml 95% ethanol)

1% trypsin solution in Insect saline (trypsin breaks the bonds holding the imaginal disc together, essentially 'releasing a coiled spring')

Protocol for ecdysone treatment

1. Sterilize instruments in 70% ethanol for 10 minutes.
2. Prepare your four Petri dishes by filling each with 2 ml of serum-supplemented Schneider's Insect Medium, then:

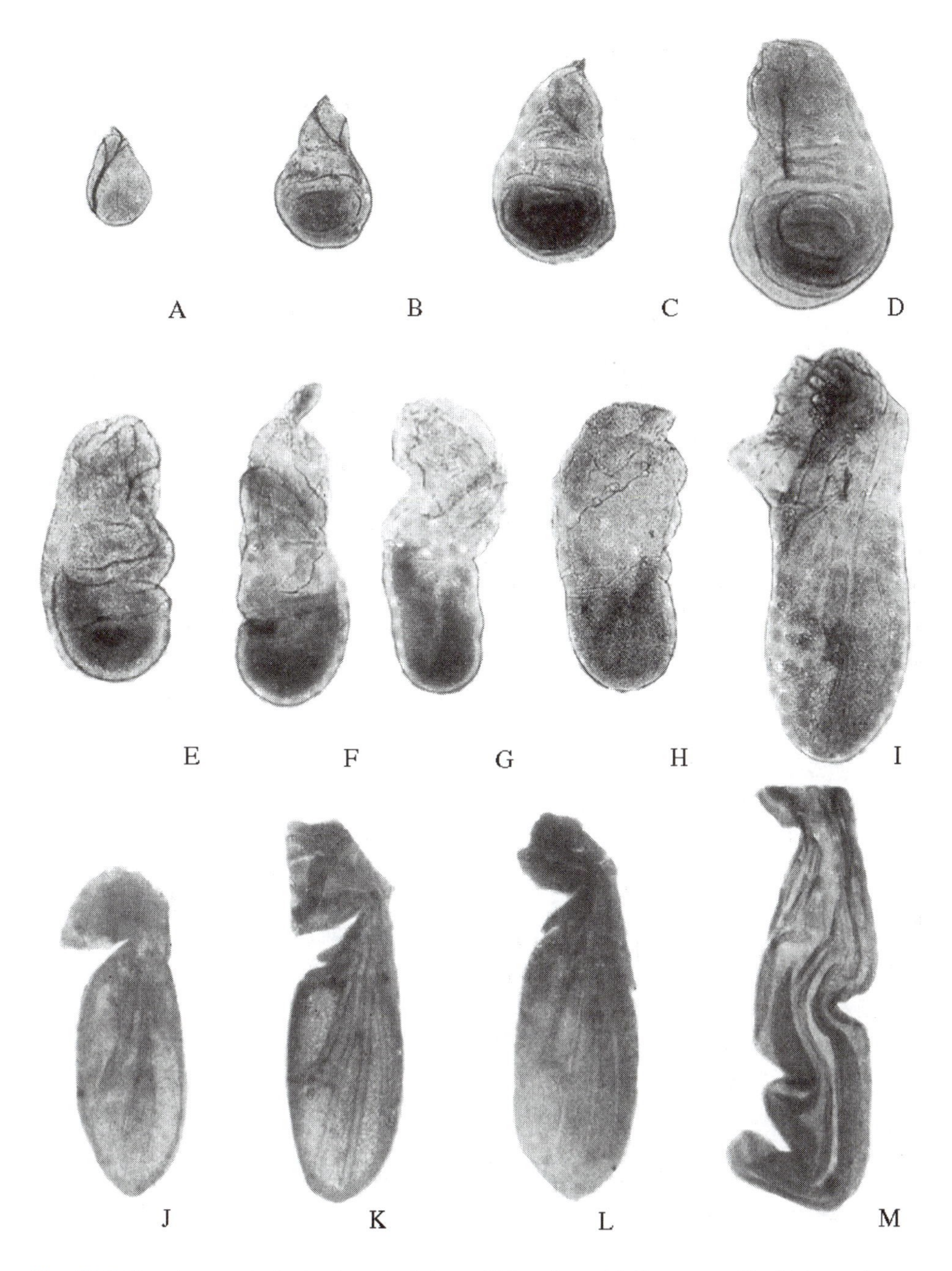

Fig. 12.3 Development of wing imaginal discs. (Taken from M. Demerec, *The biology of* Drosophila, 1950)

(a) Add 5 μl ecdysone solution to the first dish;

(b) Add 2 μl ecdysone solution to the second dish;

(c) Add 1 μl ecdysone solution to the third dish;

(d) Add 5 μl 95% ethanol to your fourth dish as a control.

Be sure to label each dish! Don't forget to add your name!

3. Obtain a 3rd-instar larva, place it on a clean slide, and rinse any debris off with Insect saline.
4. Transfer the larva to another sterile Petri dish and dissect out as many imaginal discs as you can. Grasp the head end of your larva just behind the mouth parts with one pair of forceps. Grasp the middle of the larva with the other pair and gently pull the body away from the head. The imaginal discs will be concentrated in the anterior third of the body, along the midline and attached to the tracheoles or gut.
5. Very gently transfer a few discs into each of the prepared solutions using the microneedles.
6. Sketch each disc and try to identify which discs you have.
7. Observe the discs every 30 minutes for 2 hours, looking for the beginning of eversion.
8. Allow the discs to incubate for 24 hours at room temperature.
9. Sketch and describe the discs.

Protocol for trypsin treatment

1. Place 2 ml of the 1% trypsin solution in a sterile Petri dish.
2. Transfer several imaginal discs into the prepared dish.
3. Observe the discs under a dissecting microscope for 60 minutes. Did the discs evaginate?
4. Transfer an evaginated disc to a depression slide, view under the compound microscope and sketch.

QUESTIONS

1. Did all the discs telescope out?
2. Were some discs more sensitive to the ecdysone? to the trypsin?
3. How do the ecdysone and trypsin treatments compare?

EXTERNAL INFLUENCES ON DEVELOPMENT

13

Amphibians, chicks, or fish (design your own experiment)

1 month

We have all heard about 6-legged frogs and other malformed amphibians and fish. Some of these malformations are thought to be caused by things like ultraviolet radiation, parasites, or pollution. In this lab session you will have the opportunity to see for yourself what can happen to animals living in an environment that isn't quite toxic or dangerous enough to kill them, but that can cause malformations. You need to think of some mutagenic agent to which you can expose a batch of embryos. It can be anything from alcohol (e.g. fetal alcohol syndrome), UV radiation, drugs, or water samples from interesting sources (drainage pipes, retention ponds, marinas, lakes, etc.). Are there any agricultural operations nearby? There may be pesticides, herbicides, or fertilizers that run off into streams or lakes and impact amphibian or fish populations. Industrial effluents and residential lawn and pool run-off may also be having deleterious effects on local aquatic organisms. Be creative; the only limitation is that your polluting agent must be something any embryo might normally encounter in a natural environment. The concentration you choose to expose your embryo to should correlate (to some degree) with the amounts of that chemical your organisms might be exposed to in real life. Material safety data sheets (MSDS) will be helpful in ascertaining the potential impacts of the chemicals.

You may also be interested in exploring the effects of common over-the-counter drugs on developing embryos. Many of these drugs carry warnings to pregnant mothers, but don't provide any details. Is there really a difference between the effects of aspirin and Tylenol? Does it really matter if you double-dose yourself? The concentrations of the drug you choose to expose your embryos to should correlate with the concentration that would normally be found in the bloodstream (divide the recommended dosage by the blood volume of the average male or female). You will also be able to research the drug's mode of action in a physician's desk reference (PDR) book, which may also give some indication of the types of problems (developmental or otherwise) that have been observed.

Protocol

1. Choose a potential mutagen.
2. Choose 3–5 concentrations of this mutagen to expose your embryos to. Be sure the concentration ranges from something that might have no effect on the embryos to one that could very well kill the embryo; this will allow you to see all

the potential effects of that mutagen. A good starting point is to choose a concentration that would normally be used (recommended dose of a drug or proper use of an agricultural chemical, you also can use the scientific literature to find out what concentrations of pollutants are normally found in industrial waste) and then reduce and increase that concentration by factors of 10 to bracket the original concentration. You may need to refine this range of concentrations after your first set of experiments. Be sure to include a control!

3. If you are exposing embryos to an environmental pollutant, leave the jelly coats in place, since the embryos would be protected by that jelly coat in their natural environment. If, on the other hand, you are using the embryos as a model system to explore the potential impacts of drugs on a developing human embryo, you should dejelly the embryo.

4. Document the appearance of your embryos before, during, and after their exposure to the mutagen and compare to your control animals. Be sure to consider the appearance of the jelly coat (thickness, consistency, opacity, etc.) if you have left it in place.

5. After 1 week, you will be able to determine if your chosen concentrations are too high or too low. Once you have settled on a realistic set of treatment concentrations, repeat two more times.

6. Analysis of your data will be qualitative, as this sort of analysis does not lend itself to statistical examination.

QUESTIONS

1. How did your experimental animals compare to the controls?
2. What were the major effects of the mutagen?
3. Were you able to observe a dose-dependent response to the mutagen?
4. Was there a correlation between the way your experimental animal responded to the mutagen and the way humans or other organisms have responded (check the literature)?

PART 2

Information for the Instructor

APPENDIX 1

Animal Care and Maintenance

Axolotl

The Indiana University Axolotl Colony needs to receive your order a week in advance. They ship (FedEx overnight) on Mondays and Tuesdays for deliveries on Tuesday or Wednesday respectively. If you don't need the axolotls right away, they can be stored in the spring water they arrive in, in a large finger bowl at 4°C (refrigerator). Although axolotls tolerate the cold temperatures well, they shouldn't be stored for more than a week for optimal results.

It is extremely difficult to rear axolotls to metamorphosis; they generally die 2–3 weeks post hatching, due to difficulties feeding. The Axolotl Colony does sell older, metamorphosed animals if needed.

Chick

Chicken eggs should be ordered two weeks prior to your lab to insure an appropriate arrival date. You may store the eggs at room temperature for a few days to delay development before placing them in the incubator. They should be stored in a humidified incubator at 37.5 °C. If your incubator doesn't have a humidifier, keep a large finger bowl of water inside to provide moisture. You can expect 70% of eggs to be viable (contain embryos), although this can vary seasonally (lower viability in the fall). As you place eggs in the incubator (on their sides), mark them with a pencil; this will help you keep track of their orientation during development. Eggs should be turned once or twice a day. If you choose to incubate the eggs to hatching, do not turn after the 18th day. Chicks should hatch at 21 days.

Drosphila

Order flies to arrive about two weeks prior to your lab to allow time for a large supply to develop. When the fly medium in the vials starts getting low or looking nasty, transfer flies to new vials. New vials can be purchased from Ward's or Carolina. Prepare a commercially available fly medium, add about 1 inch depth to a vial with a sprinkle of baking yeast and add a plastic net for the flies to land on. Prepare to transfer flies by tapping the vial firmly to knock adults to the bottom of the vial. Quickly invert the vial over the new vial and tap firmly to knock flies into the new vial. Separate the two vials and quickly stopper with foam plugs. If you are having difficulties removing the adults from the vial, they can be slowed down by placing the vial in the freezer for 10–15 minutes prior to shaking them out. You will be most interested in 3rd instar larvae; these large larvae will be clearly visible, climbing up the sides of the vial.

To obtain good embryos, prepare soft agar (~1%) with grape juice, and coat several slides with a 2–3 mm layer. Cut a few grooves in the agar and paint with a slurry of yeast & water. Place slides inside a culture vial, add flies and wait a few hours. Eggs will be laid on the surface and can be rinsed off with insect saline.

Lumbriculus

Store in a large finger bowl of spring water under several strips of lab paper towels (brown, unbleached are best). Exchange about half of the spring water every few days. They can be fed sinking fish pellets every few days, after which the water and paper towels should be completely replaced, however, don't feed worms during the week prior to your regeneration study. Worms can be sacrificed with an overdose of MS-222 when no longer needed.

Planaria

Several species of Planaria are available from most biological supply houses; try to order the largest specimens possible. Store in a small finger bowl of spring water in a cool, dark place. They can be fed pieces of beef liver (leave in their tank for a few hours), after which their water should be completely replaced to prevent fouling of the water. Do not feed during the week before the experiment or while regenerating. Planaria can be sacrificed with an overdose of MS-222 when no longer needed.

Sea urchin

Due to the difficulties in controlling urchin spawning, order urchins so they will arrive on the day of your lab. You may have some difficulties here with bigger commercial operations. Your best bet is to find a smaller, more local dealer. Prepare about 5 gallons of Instant Ocean (@1.021–1.025 ρ) in a 10 gallon aquarium. Check on the urchins as soon as they arrive for signs of spawning. Gametes will remain viable for several hours after spawning (hopefully long enough for your lab to begin).

If the urchins do not spawn in response to the potassium chloride or electric current, you should be prepared to cut them open to extract the gametes. Use a pair of lab scissors & cut around the urchin's equator. Testes will be white and may be relatively small. Viable ovaries, which will fill large areas of the body cavity, will be a dark reddish-brown. Urchins can be sacrificed with hypothermia, by placing in a freezer for several hours.

Zebrafish

Purchase locally or through Carolina or Ward's and set up in a 10 gallon tank with a Whisper Filter, air stone and timer-controlled lighting system. Zebrafish are kind of hyper, so provide them with some cover (fake plants). You should have at least 12 fish in a tank. They will spawn at dawn (or within 1½ hours of that) when placed on a 14L:10D cycle. You can easily adjust their normal pattern by enclosing the tank in black plastic (I use a PVC frame underneath) and gradually shifting their 'dawn' by ½ hour per day until it is set to 1–2 hours ahead of your lab. Two days before you need embryos, you should give the fish a 14 hour day and 2 small finger bowls of clear glass marbles to lay eggs in (if there aren't any marbles to protect the eggs, the adults will eat them all). Zebrafish will eat fish flakes and/or brine shrimp. These common aquarium fish are easy to maintain and should be maintained from year to year. If there is some need to sacrifice them, an overdose of MS-222 will do the trick.

APPENDIX 2

Animal and Plant Sources

Major suppliers:

Carolina Biological (**www.carolina.com**)

Fisher Scientific (**www.fishersci.com**)

Sigma-Aldrich Chemicals (**www.sigma-aldrich.com**)

Ward's Scientific (**www.wardsci.com**)

Amphibian embryos

Axolotl (*Ambystoma mexicanum*) embryos (including albinos)

Carolina: January–March

Indiana University Axolotl Colony (Jordan Hall 142, 1001 E. 3rd St.,

Bloomington, IN 47405, axolotl@indiana.edu): all year

Bullfrog—Carolina available April–August

Frog—Carolina, Wards,

Frog, toad & salamander—Charles Sullivan Co. (6685 Holt Road, Nashville, TN 37211, (615)-832-0958, **sullivan@bellsouth.net**): Most frog, toad & salamander eggs are available in late winter & spring.

Toad—Wards

You may also collect amphibian embryos in the field during the rainy season. In many places, frogs will spawn after a heavy rainfall. Collect eggs first thing in the morning, as they tend to develop quickly in warm summer temperatures.

Bush bean

seasonally (late winter—summer) from garden supply stores

Bridal veil (*Gibasis geniculata*)

special order from local florist

Chicken eggs (*Gallus domesticus*)

Carolina, Wards or local breeders (check with your local Ag Extension)

Drosophila virilis

Carolina, Wards

Killifish cysts

Wards

Oligochaete worms (*Lumbriculus*)

Carolina

Marsilea sporocarps

Carolina, Wards

As many as 25% of the sporocarps may be duds, so order 2 sets for a class of 15.

Planaria

Dugesia dorotocephalus

Carolina, August–February

Dugesia tigrina

Carolina, Wards

other species available from Carolina

Sea urchins

Gulf Specimen Marine Laboratories, Inc. (Panacea FL, **www.gulfspecimen.org**), Carolina, Wards. Purple sea urchins (*Arbacia punctulata*) are fertile between October and April in the Gulf of Mexico, while short spined urchins (*Lytechinus variegatus*) are fertile between May and September.

Spiderwort (*Tradescantia*)

wild (bloom in spring), Carolina, Wards

Sponges

saltwater aquarium shops are the simplest & cheapest source

Sweetpea

special order from local florist

Zebrafish (*Danio rerio*)

local pet store, Carolina, Wards

APPENDIX 3

Chemicals Sources

Major suppliers:

Carolina Biological (**www.carolina.com**)

Fisher Scientific (**www.fishersci.com**)

Sigma-Aldrich (**www.sigma-aldrich.com**)

Product

Actinomycin D—cell culture tested (A9415 Sigma)

Carrot Callus Initiation Medium (RG-19-3990 Carolina)

Carrot Shoot Development Medium (RG-19-4000 Carolina)

Cycloheximide Ready made (C4859 Sigma)

Cytochalasin B (C6762 Sigma)

DMSO (Dimethylsulfoxide) (D136-1 Fisher)

Ecdysone (20-Hydroxyecdysone) (H5142 Sigma)

Fetal bovine serum, insect cell culture tested (F0643 Sigma)

Formalin (37% solution) (HT50-1-2 Sigma)

Instant Ocean (local pet store)

MS-222 (Tricaine, or ethyl 3-aminobenzoate, methanesulfonate salt) (A5040 Sigma)

Not-1 primary antibody (15.3B9, chick origin, supernatant from Jessel & Dodd, University of Iowa Developmental Studies Hybridoma Bank, **www.uiowa.edu/~dshbwww/**)

Penicillin-G Potassium, embryo tested (BP914-100 Fisher)

Plant tissue quality agar (RG-19-8200 Carolina)

Retinoic Acid (all-trans retinoic acid) (R2625 Sigma)

Schneider's Insect medium, insect cell culture tested (S9895 Sigma)

Streptomycin sulfate (BP910-50 Fisher)

Trypsin tablets from porcine pancreas (T7168 Sigma)

APPENDIX 4

Slides Needed and Sources

Major suppliers

Connecticut Valley Biological: 1–800–628–7748; 82 Valley Rd, PO Box 326, Southampton, MA 01073

Triarch: 1–800–848–0810; PO Box 98, Ripon, WI 54971–0098

Ward's Scientific: **www.wardsci.com**

Product

Chick wholemount: 18, 24, 36, 48, 72, and 96 h (Triarch), also sectioned embryos

Frog embryos: sectioned blastulae, gastrulae, neurulae, 4- and 7-mm embryos (Triarch)

Germinating pollen (Ward's Scientific)

Lilium double fertilization (Ward's Scientific)

Lilium megaspore mothercell (Ward's Scientific)

Lilium ovule (Ward's Scientific)

Lilium young embryo with endosperm (Ward's Scientific)

Mixed pollen (Ward's Scientific)

Ovaries: human (Triarch), cat, frog (Triarch), chick (Connecticut Valley)

Sea urchin: all stages (Ward's Scientific)

Sperm: human (Triarch), cat, frog (Triarch), chick (Connecticut Valley)

Testes: human (Triarch), cat, frog (Triarch), chick (Connecticut Valley)

APPENDIX 5

Solutions and Other Materials

Aceto-orcein stain

1. Dissolve 6 g orcein powder in a hot mixture of:
   - 150 ml glacial acetic acid
   - 60 ml lactic acid
   - 90 ml deionized water (dH_2O)
2. Filter twice and store (indefinitely) at 4 °C.

Carrot Callus Initiation Medium

Available in powdered form from Carolina Biological; or prepare Gamborg, Miller, and Ojima Salt Base Medium as follows:

$(NH_4)_2SO_4$	134 mg
H_3BO_3	3 mg
$CaCl_2{\cdot}2H_2O$	150 mg
$CoCl_2{\cdot}6H_2O$	0.025 mg
$CuSO_4{\cdot}5H_2O$	0.025 mg
$FeSO_4{\cdot}7H_2O$	27.8 mg
$MgSO_4{\cdot}7H_2O$	250 mg
$MnSO_4{\cdot}H_2O$	10 mg
KI	0.75 mg
KNO_3	2.5 g
Na_2EDTA	37.3 mg
$Na_2MoO_4{\cdot}2H_2O$	0.25 mg
$NaH_2PO_4{\cdot}H_2O$	150 mg
$ZnSO_4{\cdot}7H_2O$	2 mg

Thiamine HCl	10 mg
Pyridoxine HCl	1 mg
Nicotinic acid	1 mg
i-Inositol	100 mg
2,4-D (synth. auxin)	1 mg
dH_2O	1000 ml (you may not need all this)
Sucrose	20 g
Agar (plant tissue qual.)	8 g

1. Add above ingredients (except last three) to 800 ml distilled/deionized water and mix until dissolved.
2. Add 20 g sucrose and dissolve.
3. Adjust pH to 5.7 with 1 mol/l NaOH or 1 mol/l HCl.
4. Bring the volume to 1 L.
5. Add 8 g agar* and heat to dissolve.
6. Autoclave for 15 minutes (121 °C) and dispense into culture dishes under sterile conditions.

***Note**: The agar concentration is important as it can impact development if too hard or too soft.

Carrot Shoot Development Medium (1 L)

Available in powdered form from Carolina Biological; or prepare Gamborg, Miller, and Ojima Salt Base Medium as for Carrot Callus Initiation Medium **but** replace the 1 mg 2,4-D with 0.2 mg kinetin.

Chick saline (123 mmol/l)

0.72% NaCl in dH_2O

Cobalt chloride solution

1% solution in dH_2O

DAB/PBT stock

1 DAB (3,3′-diaminobenzidine) tablet, or 10 mg powder, in 10 ml PBT

Ecdysone solution

20-Hydroxyecdysone	1 mg
95% ethanol	1 ml

Embryo water (a.k.a. artificial spring water)

Stock

$CaSO_4$	19 g
$MgSO_4$	1.9 g
KCl	1.5 g
NaH_2PO_4	0.4 g

Make up to 1 L with dH_2O.

Working solution

Use 10.5 ml stock in 4 L dH_2O.

Fern medium

NH_4NO_3	0.5 g
KH_2PO_4	0.2 g
$MgSO_4 \cdot 7H_2O$	0.2 g
$CaCl_2 \cdot 2H_2O$	0.1 g
Ferric citrate	5 ml of a 0.1% solution (made by dissolving 0.1 g in 100 ml boiling H_2O)

1. Make up to 1 L with dH_2O.
2. Autoclave to sterilize, and store at 4 °C.

4% Formaldehyde in 0.2 mol/l phosphate buffer

1. Make up buffer as follows:
   - Solution 1: 27.6 g $NaH_2PO_4 \cdot H_2O$ + 1 L dH_2O (store solution in refrigerator);
   - Solution 2: 28.4 g Na_2HPO_4 + 1 L dH_2O (store solution in refrigerator);
   - Mix 28 ml of Solution 1 with 72 ml of Solution 2, adjust pH to 7.4.
2. Add 10.8 ml 37% formalin solution, mix, and store in the refrigerator for up to 1 week (a dark bottle isn't necessary).

Formalin solution (0.01% for use as fungicide)

Stock (2.5%)

6.7 ml 37% formalin in embryo water

Working solution

Dilute 1 ml stock to 250 ml with embryo water.

Glass bridges

1. Cut coverslips with a diamond pencil into 4 × 12 mm pieces.
2. Pass edges slowly through a flame to smooth the edges.
3. Bend each piece by holding it over the flame at an angle until it bends under its own weight.
4. Make several shapes (angles).

4% Glutaraldehyde in 0.1 mol/l PBS

Warning: Glutaraldehyde is corrosive and can cause skin and respiratory irritation, handle with care and dispose of properly (see Chapter 2).

Insect saline

NaCl	1.87 g
KCl	0.875 g
$CaCl_2$	0.0078 g

Make up to 250 ml with dH_2O.

MS-222

This anesthetic is commonly used for aquatic or marine organisms. To sacrifice an animal, siphon off as much water out of the holding chamber (e.g. Petri dish) as possible without stressing the animal. Shake out a small amount of MS-222 into the chamber and mix. The animals may zip around the chamber for a second or two, but will rapidly become unconscious and die. If the animals do not stop moving, add a little more MS-222.

Neutral Red-stained agar chips

1. Prepare a 1–2% agar solution (powder in dH_2O) and heat to boiling.
2. Pour the agar onto clean microscope slides (as much as agar as surface tension will hold).
3. Allow the slides to dry for 2 days.
4. Place the slides in a 1% Neutral Red solution for 1–2 days.
5. Briefly rinse the slides in dH_2O and allow them to dry for several days. Once dry, slides may be stored indefinitely in a Petri dish, a jar, or anywhere dry.
6. To use, scrape strips of agar off a slide (it may help to moisten the agar before scraping) and cut them into small chips.

Nickelous ammonium sulfate ($NiNHSO_4$)

1% solution in dH_2O

PBS (phosphate buffer solution)

NaCl	8 g
KCl	0.2 g
Na_2HPO_4	1.44 g
KH_2PO_4	0.24 g

1. Make up to 1 L with dH_2O.
2. Adjust to pH 7.4.

PBT (phosphate buffer with Triton X-100)

Bovine serum albumin	0.2 g
Triton X-100	100 µl
PBS	to 100 ml

Store in the refrigerator.

PBT + N (PBT + normal goat serum)

PBT	4 ml
Normal goat serum	20 µl

Use fresh, **don't** store.

Pipettes for transferring embryos

Use disposable plastic pipettes and cut about 4.5 cm off the tip. You will be left with a large enough aperture for the embryo to move through without being damaged.

Pollen tube medium (Plain pollen tube medium)

Sucrose	100 g
H_3BO_3	0.1 g
$Ca(NO_3)\cdot 4H_2O$	0.3 g
dH_2O	800 ml

1. Mix and bring to 1 L.
2. Filter-sterilize (do **NOT** autoclave).

Pollen tube medium + actinomycin D

Stock

100 µl DMSO + 1 mg actinomycin D

Working solution

30 µl stock + 10 ml Plain pollen medium

Pollen medium + cycloheximide

Stock

1 ml DMSO + 10 mg cycloheximide

Working solution

200 µl stock + 10 ml Plain pollen medium

Pollen medium + cytochalasin B

Stock

100 µl DMSO + 0.5 mg cytochalasin B

Working solution

40 μl stock + 10 ml Plain pollen medium

Potassium chloride (0.5 mol/l KCl)

3.73 g KCl in 100 ml dH_2O

Rearing solution (Holtfreter's solution) (RS)

Make up Solutions 1–4 ahead of time and mix RS as needed.

Stock solutions

Solution 1: 1.6 mol/l NaCl (23.5 g NaCl + 227.5 ml dH_2O)

Solution 2: 0.13 mol/l KCl (2 g KCl + 198 ml dH_2O)

Solution 3: 0.09 mol/l $CaCl_2$ (2.65 g $CaCl_2 \cdot 2H_2O$ + 199 ml dH_2O)

Solution 4:

- 72.8 ml of Solution 1
- 20 ml of Solution 2
- 10 ml of Solution 3
- 900 ml dH_2O

To make 2 liters of RS

1. Mix a 1:1 solution of Solution 4 and dH_2O.
2. Sterilize by boiling for 3 minutes.
3. Cool to room temperature and add:
   (a) 400 mg penicillin G (potassium salt) (stock penicillin: 100 million unit vial contains 1600 units/mg)
   (b) 400 mg streptomycin sulfate
   (c) 200 mg $NaHCO_3$
4. Store at 4 °C.

Rearing solution—calcium-free

Follow the instructions for RS, but leave out the $CaCl_2$ (Solution 3).

Rearing solution + 1% KOH

Add 1 g KOH to 99 ml RS.

all-*trans* Retinoic acid

Warnings:

- Retinoic acid is a **teratogen**, handle with care and dispose of properly (see Chapter 2).
- DMSO will move chemicals across **your** membranes, handle with care and dispose of correctly (see Chapter 2)

1. Make a 10^{-2} mol/l solution in DMSO.
2. Take 20 μl aliquots and store in the freezer.

Seawater

Instant Ocean (follow directions on the box)

Seawater—calcium–magnesium-free

NaCl	30.7 g
KCl	0.745 g
EGTA	0.76 g
$NaHCO_3$	0.21 g

Bring to 1 L with dH_2O.

Trypsin solution (1%)

1. Dissolve 1 trypsin (porcine) tablet (1 mg) in 1 ml dH_2O (trypsin tablet is already buffered).
2. Use within 30 minutes.

Tungsten microneedles for embryo dissection

0.004-inch tungsten rods

Glass Pasteur pipettes

1. Cut off most of the glass tip of the pipette, leaving about 1 cm behind.
2. Soften the tip of the pipette in an open flame and partially compress it, leaving an opening large enough to take the tungsten rod.

3. Insert the tungsten rod and compress and melt the glass to firmly hold it in place.
4. Cut the rod to size and bend the tip to 70–90 °.

For dissection, use one microneedle to hold the embryo down, and the other to tease apart the embryo. If you wish to make more precise cuts, you may want to make a microscalpel (cut a sliver of a double-sided disposable razor blade with scissors (will be bent), clip into a small surgical clamp, and cut through the skin before using the microneedles).

Forceps

If the tips of your forceps have been bent, they may be sharpened again with polishing stones.

PART 3

Glossaries

GLOSSARY OF TERMS

aboral The surface of a metamorphosed echinoderm that is opposite to the mouth. The aboral surface contains the anus, madreporite (connects to the water vascular system), and the gonopores.

adhesion Stickiness between similar cells or material. Calcium-dependent adhesion molecules (CAMs) are crucial to morphogenetic events and the integrity of tissues.

albumin Refers to the white of a bird egg, containing the protein albumin, or to the blood plasma-protein seralbumin.

allantois Extra-embryonic membranes in bird and reptile embryos aids in respiration and excretion. In mammals, this membrane contains blood vessels that transport materials to and from the placenta.

amnion The innermost extra-embryonic membrane of birds, reptiles, and mammals that surrounds the embryo.

amoeboid cells The feeding cells of sponges (genus *Porifera*), which are also used to transport inorganic material to the osculum and can store nutrients.

angiosperm A group of plants that enclose their seeds in a seed vessel (i.e. fruit); flowering plants.

antheridium The part of the plant that produces pollen.

apical Refers to the tip of the plant, limb bud, etc.

archegonium Female reproductive organ of a plant (produces the egg).

archenteron The embryonic gut that forms as mesoderm and endoderm invaginate during gastrulation.

area pellucida In the chick embryo, this is the clear area in the center of the blastoderm, under which lies the subgerminal space.

aril The outer covering of a seed that develops from the ovule after fertilization.

blastema A cluster of cells proximal to an amputation site that will proliferate and dedifferentiate in anticipation of regenerating a traumatically lost limb.

blastocoel This is a fluid-filled cavity within a blastula.

blastopore The invagination on the surface of a gastrulating embryo that marks the point through which mesoderm and endoderm move into the embryo.

blastula An early embryo that has, through simple radial cell division, formed a hollow ball of cells.

blood islands Area in the visceral lateral plate of the developing embryo that will give rise to blood cells and vessels of the circulatory system.

bottle cells These are cells in the region of the blastopore, which, upon lengthening, initiate involution.

callus A cluster of dedifferentiating and proliferating cells in a plant or plant fragment. This is common at a site of injury.

carpel The flower organ containing the ovules.

chorion This outermost membrane in reptile, bird, and mammal embryos is used for respiration. Mammals have incorporated the chorion into the placenta and also use it for waste removal.

colloid A homogeneous, gluey substance that may contain the spores of some plants.

convergent extension During this process, a sheet of cells can become narrower and longer.

corpus luteum An endocrine body that forms around a recently ruptured ovarian follicle.

cortical granules These are located just under the plasma membrane and are released into the perivitelline space following depolarization and release of calcium. The granules absorb water, expand, and, in conjunction with the vitelline membrane, create the fertilization membrane.

cotyledon This is the first leaf of a plant that serves as a source of food for the new plant before it is able to photosynthesize.

diapause A dormant period during development, during which cellular processes are either arrested or considerably slowed. The embryo is often contained in a cyst.

diencephalon The smallest part of the vertebrate brain, containing the epithalamus, thalamus, subthalamus, and hypothalamus.

diplotene Phase during meiotic prophase, during which homologous chromosomes begin to pull apart but are still held together by the chiasmata.

ecdysone The insect hormone responsible for initiating pupal and adult metamorphosis. Environmental conditions stimulate the corpus allatum to secrete prothoracicotropic hormone, which stimulates the prothoracic gland to release ecdysone.

endosperm The tissue surrounding the embryo that serves as a food source.

epiboly This occurs during gastrulation as the ectoderm thins and spreads to cover the whole embryo.

epidermal cell These are the outer 'skin' cells of a sponge that can respond to potentially dangerous chemical or mechanical stimuli by closing off the incurrent pores.

epimorphosis The form of regeneration in which new structures are formed as the result of new growth, rather than repatterning (e.g. planaria).

epiphysis The portion of a long bone, separated from the main shaft by cartilage, where ossification takes place.

explant A piece of tissue (plant or animal) that has been removed from its source to be cultured *in vitro* or *in vivo* (in another host).

flagellated cell A type of cell in sponges (located inside the main chamber), also known as collar cells or choanocytes, that beat rhythmically to draw water into the sponge for filter feeding.

follicle cells Cells in *Drosophila* that surround the oocyte and nurse cells during early development. These cells help with the uptake of yolk proteins.

gamete In all animals and plants, the cells that will contribute to the formation of a zygote.

gametophyte Gametophytes are only in plants. In animals, the gamelocyte (spermatocyte or oocyte) gives rise to the gametes.

gastrula The stage of development during which the mesoderm and endoderm move from the surface into the blastocoel through the blastopore to form the archenteron. At the end of gastrulation, all three germ layers are in the correct location.

gonopore These are the pores, located on the aboral surface of sea urchins, through which gametes are exuded during spawning.

halteres These are balancing organs in *Drosophila*, located near the wings.

heat shock This is a technique used to cause genes to unravel and undergo transcription (creating puffs) at a time when they would normally remain tightly packaged.

Hensen's node This is a cluster of cells at the anterior end of the primitive streak in chicks that is responsible for the development of the notochord, nervous system, and somites. Hensen's node corresponds to the vertebrate Spemann organizer.

hindbrain The part of the brain that includes the pons, cerebellum, and medulla oblongata.

homeobox A region in a gene containing homeodomain motifs that code for transcription factors and are involved in patterning body axes.

homeobox gene This refers to any member of a group of genes containing a homeodomain DNA-binding motif. These genes are very common in transcription factor genes and are often highly conserved.

homeotic gene A gene, that when mutated, causes body parts to be replaced with organs or structures from other parts of the body (i.e. Antennapedia mutations can result in legs growing from the forehead, where antennae should be).

hyaline membrane The outermost, and most impenetrable, membrane formed in response to fertilization to prevent polyspermy.

hypocotyl The part of the developing plant embryo or seedling located below the cotyledon and above the radicle.

imaginal disc Epithelial sacs in insect larvae that will give rise to legs, wings, eyes, antennae, and genitalia following metamorphosis and exposure to the hormone ecdysone.

instar The life stage in between each of the molts made by a juvenile or larval insect.

intercalate This refers to the movement of two layers of neighboring cells into a single layer. Intercalation is used during epiboly when the ectoderm must thin to cover the gastrulating embryo.

involution This refers to the initial movement of cells into the blastocoel during gastrulation. The bottle cells constrict their apical ends, thereby forcing the rest of the cell into the embryo.

jelly coat The thick, protective coating surrounding the embryos of amphibians.

labial Refers to lips or any lip-like structure.

labrum A lip, edge or lip-shaped structure.

lampbrush chromosome The area of a chromosome that has 'unraveled' for the purposes of transcription, giving that area of the chromosome the appearance of a brush.

larva An immature, wingless, usually worm-like form of an insect or the post-hatching phase of an animal that must pass through metamorphosis before it resembles the adult form.

limb bud The presumptive limb, containing zones of proliferating cells and organizing centers.

luteinizing hormone A hormone secreted by the anterior pituitary gland that stimulates the female to ovulate and the male to produce testosterone.

madreporite The opening at the top of a sea urchin connected to the water vascular system and surrounded by gonadal pores.

megaspore A large plant structure that will develop into a (female) megagametophyte, which then produces eggs.

mesencephalon The part of the brainstem containing the roofplate, crus cerebri, substantia nigra, and motor nuclei of the oculomotor and trochlear nerves.

metencephalon The anterior part of the hindbrain, containing the pons and cerebellum.

micropyle This is a small, tube-like opening through which the sperm reaches the ovum.

microspore A small plant structure that will give rise to a (male) microgametophyte and then produce sperm.

midbrain Synonym for the mesencephalon.

morphallaxis The type of regeneration, typified by *Hydra*, in which existing tissues are rearranged to replace the missing structures; no new growth is required.

mouth hooks Extendible appendages that bring food into the mouth of *Drosophila* larvae.

myelincephalon Synonym for the medulla oblongata.

neuromere There are three divisions, or neuromeres, in the developing brain of *Drosophila*.

neurula Refers to the stage of development during which the nervous system is forming.

notochord A hollow, flexible rod found ventral to the nerve cord in all chordates. The notochord is used for muscle attachment, induces patterning of the neural tube, and is eventually incorporated into the vertebral bodies of vertebrates.

nurse cells Polyploid cells that are found at one end of the *Drosophila* egg follicle and produce large amounts of proteins and other vital cell products for the developing embryo.

oocytes An immature haploid ovum (primary or secondary oocytes).

oogonia The diploid cell, derived from the germ cells, from which the oocytes develop.

osculum The large opening at the top of a sponge through which filtered water is expelled.

ovule The part of a plant ovary, in the carpel, where megaspores develop and where the embryo sac is located.

paedomorphic Refers to an adult organism that retains larval or juvenile features.

paraxial mesoderm The mesoderm located on either side of the neural tube that will give rise to the somites.

photoperiod The daily cycle of light and dark hours; a long photoperiod occurs in the summer.

plumule The first true leaves of a plant.

pluteus The free-swimming larval form of echinoderms that immediately follows the prism stage and precedes metamorphosis. These larvae are pyramidal, with a long 'arm' projecting from each basal corner of the larva.

polyspermy Penetration of the egg by more than one sperm. This is not desirable, as the zygote will have too many chromosomes.

polytene This refers to the large chromosomes in the salivary glands of *Drosophila*, which are clusters of many copies of the same gene, formed by replication without mitosis.

primitive groove The first sign of neurulation in a chick, which extends, as a shallow depression, through the middle of the primitive streak.

primitive pit The anterior end of the primitive streak and groove.

primitive streak This thickened area denotes the anterior–posterior axis of the chick embryo and is found in the middle of the area pellucida.

progesterone A female hormone produced by the placenta and corpus luteum (important for ovulation).

puberty The period during which adolescents become reproductively mature, characterized by the development of secondary sex characteristics and maturation of gonads.

pupa An intermediate, quiescent stage in the development of most metamorphic insects.

radicle The nascent root of a plant embryo.

respiratory filaments Extensions of the chorion that allow gas exchange to occur with a minimum of water loss to the embryo. The filaments also allow embryos to survive immersion.

seminiferous tubules The tubules contained within the testes where spermatogenesis occurs.

somite Segmented blocks of mesoderm in vertebrates that arise from paraxial mesoderm. Somites give rise to muscles of the trunk, limb muscles, the vertebral column, and skin.

spermatocytes Immature diploid sperm that have not yet gone through meiotic division.

spermatids Immature haploid sperm that have gone through meiosis, but must still metamorphose into sperm.

spermatogonia Cells that derive from germ cells that will give rise to sperm cells.

spermatozoa Synonym for sperm.

spiracles The paired anterior and posterior spiracles are the openings to the larval *Drosophila* respiratory system (trachea).

sporocarp A structure in which spores (megaspores and/or microspores) are formed in plants.

sporophyte The plant that reproduces using asexual spores. May be one phase in the life cycle of a plant that undergoes alternation of generations.

sporopollenin A polymer found in the walls of pollen grains.

stigma The part of a flower's carpel that receives pollen (is sticky) and where pollen germinates.

style The carpel 'stalk' that connects the stigma and the ovule.

telencephalon The segment of the brain containing the cerebral hemispheres.

trachea Hollow tubes, open to the environment through the anterior and posterior spiracles, that serve as the respiratory system for larval *Drosophila*.

vitelline membrane The membrane that surrounds the egg immediately after fertilization (to block polyspermy) and protects the zygote/embryo from mechanical damage.

vitillogenesis The formation of proteins, lipids, and glycogen in the liver, their transport through the circulatory system, and accumulation in the yolk or oocyte.

zygote The cell formed after the egg and sperm nuclei fuse.

ABBREVIATIONS

14L:10D	14 hours of light:10 hours of dark
2,4-D	2,4-dichlorophenoxyacetic acid
CAMs	calcium-dependent adhesion molecules
DAB	3,3′-diaminobenzidine
dH_2O	deionized water
DMSO	dimethylsulfoxide
H_2O_2	hydrogen peroxide
Hox	homeobox gene
IgG	immunoglobulin G
MS-222	tricaine, or ethyl 3-aminobenzoate, methanesulfonate salt
MSDS	Material Safety Data Sheets
$NiNHSO_4$	nickelous ammonium sulfate
PB	phosphate buffer
PBS	phosphate buffer solution
PBT	phosphate buffer with Triton-X
PBT + N	PBT supplemented with normal goat serum
PDR	physician's desk reference
PVC	polyvinylchloride
RA	retinoic acid
RS	Rearing solution

LITERATURE CITED

Armstrong, J. B. and Malacinski, G. M. (ed.). (1989). *Developmental biology of the axolotl.* Oxford University Press, New York.

Ashburner, M. and Bonner, J. J. (1979). The induction of gene activity in *Drosophila* by heat shock. *Cell,* **17**, 241–54.

Campbell, D. H. (1905). *The structure and development of the mosses and ferns* (2nd edn). MacMillan Press, New York.

Demerec, M. (1950). *Biology of Drosophila.* Wiley, New York.

Downs, L. E. (1963). *Laboratory embryology of the chick.* Wm C. Brown, Dubuque, IA.

Downs, L. E. (1968). *Laboratory embryology of the frog.* Wm C. Brown, Dubuque, IA.

Eakin, R. M. (1958). *Vertebrate embryology.* University of California Syllabus Series No. 376. Berkeley, CA.

Freeman, W. H. and Bracegirdle, B. (1962). *An atlas of embryology.* Heinemann Press, London.

Gehring, W. J. (1998). *Master control genes in development and evolution: the homeobox story.* Yale University Press, New Haven, CT.

Hamburger, V. (1960). *A manual of experimental embryology.* The University of Chicago Press, Chicago, IL.

Hamburger, V. and Hamilton, H. L. (1951). A series of normal stages in the development of the chick embryo. *J. Morphol.*, **88**, 49–92.

Holtfreter, J. (1947). Observations on the migration, aggregation and phagocytosis of embryonic cells. *J. Morphol.*, **80**, 25–57.

Kimmel, C. B., Ballard, W. W., Kimmel, S. R., Ullmann, B., and T. F. Schilling. (1995). Stages of embryonic development of the zebrafish. *Dev. Dyn.*, **203**, 253–310.

Mauseth, J. D. (1998). *Botany. An introduction to plant biology* (2nd edn). Jones and Bartlett, Sudbury, MA. (www.jbpub.com)

Morgan, T. H. (1898). Experimental studies of the regeneration of *Planaria maculata. Arch. Entwicklungsmech.*, **7**, 364.

National Institutes of Health (1985). *Guide for the care and use of laboratory animals.* NIH Publ. No. 86–23. Bethesda, MD.

Pearse, V., Pearse, J., Buchsbaum, M., and Buchsbaum, R. (1987). *Living invertebrates.* Blackwell Scientific, Palo Alto, CA.

Roux, W. (1894). Uber das Cytotropismus der Furchungszellen des Grasfrosches (*Rana fusca*). *Arch. Ent. Mech.*, **1**, 43–68.

Schoenwolf, G. C. (2001). *Laboratory studies of vertebrate and invertebrate embryos. Guide and atlas of descriptive and experimental development* (8th edn). Prentice-Hall. Upper Saddle River, NJ.

Scott, R. J. (1995). Pollen tube formation and the central dogma of biology. In *Tested studies for laboratory teaching*, Vol. 16 (ed. C. A. Goldman). Proceedings of the 16th Workshop/Conference of the Association for Biological Laboratory Education.

Steward, F. C., Mapes, M. O., and Smith, J. (1958). Growth and organized development of cultured cells. I. Growth and division of freely suspended cells. *Am. J. Bot.*, **45**(10), 693–703.

Steward, F. C., Mapes, M. O., and Mears, K. (1958). Growth and organized development of cultured cells. II. Organization in cultures grown from freely suspended cells. *Am. J. Bot.*, **45**(10), 705–8.

Townes, P. L. and Holtfreter, J. (1955). Directed movements and selective adhesion of embryonic amphibian cells. *J. Exp. Zool.*, **128**, 53–120.

Wilson, H. V. (1907). On some phenomena of coalescence and regeneration in sponges. *J. Exp. Zool.*, **5**(2), 245–58.

INDEX

Bold numbers denote reference to illustrations